Solar and Terrestrial Radiation
METHODS AND MEASUREMENTS

To my wife Vivien

Solar and Terrestrial Radiation

METHODS AND MEASUREMENTS

Kinsell L. Coulson

Department of Meteorology
University of California
Davis, California

ACADEMIC PRESS New York San Francisco London 1975

A Subsidiary of Harcourt Brace Jovanovich, Publishers

ACADEMIC PRESS, INC.
111 Fifth Avenue, New York, New York 10003

United Kingdom Edition published by
ACADEMIC PRESS, INC. (LONDON) LTD.
24/28 Oval Road, London NW1

Library of Congress Cataloging in Publication Data

Coulson, Kinsell L
     Solar and terrestrial radiation.

     Includes bibliographies and index.
     1.   Solar radiation—Measurement.   2.   Terrestrial
radiation—Measurement.   I.   Title.
QC912.C68   551.5′271   74-17986
ISBN 0−12−192950−7

Contents

Chapter Four

Solar Radiation: Diffuse Component

Chapter Five

Ultraviolet Radiation from the Sun and Sky

Chapter Six

Illumination

Chapter Seven

Polarization of Light in the Atmosphere

Chapter Eight

Duration of Sunshine

Chapter Nine

The Solar Constant

Chapter Ten

Terrestrial Radiation: Field Characteristics

Chapter Eleven

Terrestrial Radiation: Methods of Measurement

Preface

Dr. Andrew J. Drummond was to have coauthored this book with me. His tragic illness and subsequent death prevented his participation beyond the extent of his reading and commenting on the sections of Chapters 3 and 4 dealing with pyrheliometers and pyranometers. Because of his daily contact with these instruments, I am fortunate to have had his expert advice on these sections. But I also feel certain that the entire book could similarly have profited from his counsel, and I know that I have missed the anticipated personal contacts with a good friend and colleague.

In the general design of the book, I was guided by the thought that this is an appropriate time for a summary dealing primarily with radiation instrumentation for the meteorologist or atmospheric physicist. The developments over the last decade or two have brought instruments for routine measurements of solar and terrestrial radiation at the earth's surface to a stage that is probably adequate for most meteorological and climatological purposes. The instruments, when properly used, are certainly adequate to erase many of the present deficiencies in climatological data and routine monitoring of the radiative regime of the surface and lower atmosphere. They do not, however, provide the precision required for studies of climatic change, spectral distribution of atmospheric radiation, and certain other meteorological or technological requirements. An entirely new generation of radiation instruments has been developed in the space program for use on meteorological and other types of satellites. Unfortunately these very sophisticated types of devices are beyond the scope of this book, but they probably indicate the direction that future routine monitoring devices will take.

The major emphasis in the book is on radiation instrumentation. I have tried to include enough underlying theory to make the book useful in understanding the basic radiative processes in the atmosphere and a sufficient number of historical notes to indicate the rich traditions in science to which the radiation meteorologist is beneficiary. The level of the discussions is designed for the upper division or beginning graduate college student and the professional meteorologist. The format of each chapter is to include the theory and background information in the first part and discussions of individual instruments in the last part. This did not seem feasible, however, for terrestrial radiation, so the theory and background information is included in Chapter 10 and instrumentation in Chapter 11.

I want to express my appreciation to my colleagues for furnishing information I requested, and especially to the firms and institutions for supplying, mostly without compensation, many of the instrument photographs which appear in the various chapters.

Kinsell L. Coulson

Solar and Terrestrial Radiation

METHODS AND MEASUREMENTS

CHAPTER ONE

Principles of Radiation Instruments

1.1 DEFINITIONS, TERMINOLOGY, AND UNITS

There is a great diversity in notations and terminology used in discussions of radiation so it is important at the outset to define the main quantities and terms with which we will be concerned in this book. The notation introduced by Chandrasekhar (1950), which has been widely accepted in the field, will be followed as closely as feasible in mathematical expressions, and the individual symbols are defined in Appendix A. The terminology for the various radiation fluxes and the classification of the associated instruments are those recommended by the World Meteorological Organization (1969). The units have been chosen on the basis of standard use (principally in the field of meteorology) and ease of physical interpretation. They do not necessarily conform to any one of the many sets of units which have been recommended by various organizations.

1.1.1 Solid Angle

The concept of a solid angle can be illustrated as follows. We assume a line through point 0 moving in space and intersecting an arbitrary surface located at some distance s from point 0. If the locus of the point of intersection forms a closed path on the surface but does not intersect itself, then a unique area is defined on the surface. We assume the area is an elemental area, da, the surface normal of which makes an angle γ with the direction to point 0. Then, the projected area as seen from point 0 is

1

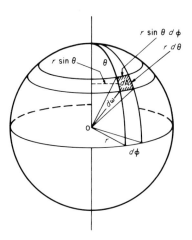

Fig. 1.1 Illustration of solid angle and its representation in spherical coordinates.

$dA = da \cos \gamma$, and the elemental solid angle, subtended at 0 by da is defined by $d\omega = dA/s^2$. Obviously for a finite area the total solid angle $\omega = \int d\omega_i$, where $d\omega_i$ is the solid angle subtended by the ith areal element dA_i.

For purposes of illustration, we assume the surface is the surface of a sphere of radius r, as shown in Fig. 1.1. For this case, $\gamma = 0$ everywhere, and the solid angle subtended at the center of the sphere by area dA on its surface is $d\omega = dA/r^2$. For the special case of a unit sphere ($r = 1$), $d\omega$ and dA have the same numerical value if $d\omega$ is expressed in steradians (sr), and dA and r are expressed in the same system of units. Since the area of the surface of a sphere is $4\pi r^2$, the total solid angle subtended at a point by the entire surrounding sphere is $4\pi r^2/r^2 = 4\pi$ sr.

A hemispheric solid angle is 2π sr. As can be seen from Fig. 1.1, an elemental solid angle is conveniently expressed in a spherical (θ, ϕ) coordinate system as

$$d\omega = \frac{(r\, d\theta)\ (r \sin \theta\, d\phi)}{r^2} \tag{1.1}$$

1.1.2 Intensity and Flux

Let us consider a pencil of radiation crossing the elemental area $d\sigma$ of Fig. 1.2 and confined to the elemental angle $d\omega$, which is oriented at some angle θ to the normal of $d\sigma$. The energy dE_ν contained in the frequency interval $d\nu$ which crosses $d\sigma$ in time increment dt is given by

$$dE_\nu = I_\nu\, d\nu\, dt\, d\omega\, d\sigma \cos \theta \tag{1.2}$$

This relation defines the monochromatic specific intensity in the most general way as

$$I_\nu = \frac{dE_\nu}{d\nu \, dt \, d\omega \, d\sigma \, \cos\theta} \tag{1.3}$$

Thus the definition of specific intensity, or simply intensity, implies a directionality in the radiation stream—an intensity in a given direction. The term flux, however, is simply a flow of energy, and it may or may not have an implied direction. For instance, the monochromatic flux of energy across $d\sigma$ is given by the integration of the normal component of I_ν over the entire spherical solid angle Ω. Thus

$$F_\nu = d\nu \, dt \, d\sigma \int_\Omega I_\nu(\omega) \, \cos\theta \, d\omega \tag{1.4}$$

or, in terms of spherical coordinates θ and ϕ,

$$F_\nu = d\nu \, dt \, d\sigma \int_0^{2\pi} \int_0^\pi I_\nu(\theta, \phi) \, \sin\theta \, \cos\theta \, d\theta \, d\phi \tag{1.5}$$

On the other hand, for radiation instrumentation purposes we are mainly interested in the radiant energy which is incident on a surface from less than the entire possible solid angle. The monochromatic flux on a plane (one-sided) surface, for instance the sensing surface of a pyranometer, is received from one hemisphere only, and is given by

$$F_\nu = d\nu \, dt \, d\sigma \int_0^{2\pi} \int_0^{\pi/2} I_\nu(\theta, \phi) \, \sin\theta \, \cos\theta \, d\theta \, d\phi \tag{1.6}$$

Then the entire flux of energy at all frequencies on a plane (one-sided)

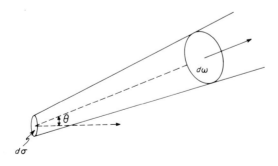

Fig. 1.2 Diagram of a pencil of radiation through elemental area $d\sigma$ and confined to elemental solid angle $d\omega$.

surface, per unit area and unit time, is

$$F = \int_0^\infty \int_0^{2\pi} \int_0^{\pi/2} I(\nu; \theta, \phi) \sin\theta \cos\theta \, d\theta \, d\phi \, d\nu \qquad (1.7)$$

Two singularities in the above relations have important implications in instrumentation. First, it is seen from the definition of intensity in Eq. (1.3) that for a finite amount of energy dE_ν the intensity $I_\nu \to \infty$ if $d\omega \to 0$. Thus the concept of intensity breaks down for parallel radiation. In that case we speak only of flux of energy from the specified direction. The second singularity occurs for the case in which the intensity is isotropic (the same in all directions). For an isotropic intensity distribution I_ν can be taken outside the integral of Eq. (1.6), which simplifies to

$$F_\nu = \pi I_\nu \, d\nu \, dt \, d\sigma \qquad (1.8)$$

and Eq. (1.7) becomes

$$F = \pi \int_0^\infty I_\nu \, d\nu \qquad (1.9)$$

1.1.3 Transmission, Absorption, Emission, and Scattering

The transfer of radiation through a medium such as the Earth's atmosphere has been discussed at length by many authors (e.g., Chandrasekhar, 1950; Goody, 1964; Kondratyev, 1969; Ambartsumian, 1958), although much still remains to be learned about the interaction of radiation with matter. In general, the medium may exhibit both absorption and scattering of incident radiation, each of which affects the characteristics of the transmitted radiation, and any real medium also emits radiation. It is well at the outset to put these concepts on a firm physical basis through the equation of radiative transfer.

The equation of transfer for a medium which absorbs, emits, and scatters radiation may be derived as follows (Chandrasekhar, 1950). We consider a pencil of radiation of intensity I_ν incident on an elemental surface $d\sigma$ in unit time, as shown in Fig. 1.3. For simplicity we consider the elemental surfaces to be normal to the incident beam. During its traverse of the pathlength ds through the medium, the incident energy is attenuated by scattering and absorption, the intensity of radiation finally emerging from the cylinder being $I_\nu + dI_\nu$. The mass of material dm of density ρ, contained in a cylinder of unit cross section and elemental solid angle, is

$$dm = \rho \, ds$$

so we can define a mass attenuation coefficient κ such that

$$dI_\nu = - \kappa I_\nu \rho \, ds \qquad (1.10)$$

In general, the attenuation coefficient is the sum of a scattering coefficient $\kappa_\nu{}^s$ and an absorption coefficient $\kappa_\nu{}^a$. The energy lost by absorption goes to heating the medium, to producing photochemical reactions, or to some other form of energy, and is thus lost to the radiation field. However, that lost by pure scattering emerges as radiation, still of the same wavelength, only the direction of propagation having been changed in the process of attenuation. This scattered component is responsible for the skylight in the case of the sunlit sky.

On integration of Eq. (1.10) over the pathlength from one arbitrary point p_1 to another p_2 in the medium, we obtain the intensity of the emergent radiation as

$$I_\nu = I_{\nu 0} \exp\left(-\int_{p_1}^{p_2} \kappa_\nu \rho \, ds\right) = I_{\nu 0} \exp[-T_\nu(p_1, p_2)] \qquad (1.11)$$

where the exponent defines the optical thickness T between p_1 and p_2. This is one form of the Bouguer–Lambert law for the transmission of radiation.

Since the scattered energy is not lost to the radiation field we can consider $\kappa_\nu{}^s$ to be also a (virtual) emission coefficient. The intensity of

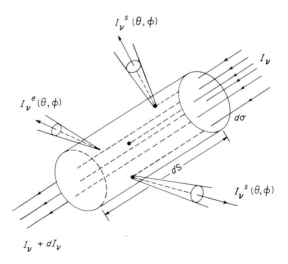

Fig. 1.3 Diagram of the change of radiant intensity over a pathlength ds in a medium which exhibits scattering, absorption, and emission.

energy $I_\nu{}^s(\theta, \phi)$ emerging from the cylinder in direction (θ, ϕ) because of scattering through pathlength ds is

$$dI_\nu{}^s(\theta, \phi) = \kappa_\nu{}^s p(\theta, \phi) I_\nu \rho \, ds \qquad (1.12)$$

where $p(\theta, \phi)$ is the phase function which describes the angular distribution of scattered radiation with respect to the direction of the original incident radiation.

The medium may also exhibit true emission of radiation. The intensity $I_\nu{}^e(\theta, \phi)$ of this emitted radiation at some angle (θ, ϕ) is given in terms of a mass emission coefficient j_ν as

$$I_\nu{}^e(\theta, \phi) = j_\nu \, dm = j_\nu \rho \, ds \qquad (1.13)$$

for a cylinder of unit cross section. The relationship between the absorption and emission coefficients and the form of the source function are discussed in Section 1.2—Radiation Laws.

1.1.4 Terminology in Radiation Instruments

Considerable confusion in the terminology applied to various types of solar radiation instruments has been built up over the years, with a given type of instrument being designated by different names and the same name being applied to different types of instruments. For instance, an instrument measuring the total hemispherical flux of solar radiation has been variously called a pyrheliometer, pyranometer, solarimeter, actinograph, and sunshine receiver or sunshine recorder.

A narrow angle instrument which measures mainly direct solar radiation has generally been termed a pyrheliometer, normal-incidence pyrheliometer, or actinometer. When either type is connected to a recorder, the suffix "meter" is often replaced by "graph," thus making pyrheliograph, solarigraph, actinograph, and so on. In an effort to standardize terminology, the World Meteorological Organization (1965), through its Commission for Instruments and Methods of Observation, has recommended the following classification of radiation instruments:

Pyrheliometer—an instrument for measuring "the intensity of direct solar radiation at normal incidence"

Pyranometer—an instrument for measuring "the solar radiation received from the whole hemisphere. It is suitable for the measurement of the global or sky radiation." A pyranometer for measuring the radiation on a spherical surface is a "spherical pyranometer"

Pyrgeometer—an instrument for measuring "the net atmospheric radiation on a horizontal upward-facing black surface at ambient air temperature"

Pyrradiometer—an instrument for measuring "both solar and terrestrial radiation (total radiation)"

Net Pyrradiometer—an instrument for measuring "the net flux downward and upward total (solar, terrestrial surface, and atmospheric) radiation through a horizontal surface." A net radiometer is sometimes termed a balance pyrradiometer or radiation balance meter

1.1.5 Radiation Units

The following are the principal units which will be used in this book. The actual selection among the units will depend on established use and convenience.

Quantities and units	Equivalent in cgs system
Wavelength	
micrometer (μm)	10^{-4} cm
angstrom (Å)	10^{-8} cm
Frequency (ν)	
sec^{-1}	sec^{-1}
Wave number	
cm^{-1}	cm^{-1}
Specific intensity : spectral	
cal cm^{-2} sec^{-1} μm^{-1} sr^{-1}	4.19×10^{11} erg cm^{-2} sec^{-1} cm^{-1} sr^{-1}
W cm^{-2} sr^{-1} cm^{-1}	10^{7} erg cm^{-2} sec^{-1} cm^{-1} sr^{-1}
Specific intensity: total	
cal cm^{-2} sec^{-1} sr^{-1}	4.19×10^{7} erg cm^{-2} sec^{-1} sr^{-1}
W cm^{-2} sr^{-1}	10^{7} erg cm^{-2} sec^{-1} sr^{-1}
Radiant flux: spectral	
cal cm^{-2} min^{-1} μm^{-1}	6.98×10^{9} erg cm^{-2} sec^{-1} cm^{-1}
ly (langley) min^{-1} μm^{-1}	6.98×10^{9} erg cm^{-2} sec^{-1} cm^{-1}
W cm^{-2} μm^{-1}	10^{11} erg cm^{-2} sec^{-1} cm^{-1}
Radiant flux: total	
cal cm^{-2} min^{-1}	6.98×10^{5} erg cm^{-2} sec^{-1}
ly (langley) min^{-1}	6.98×10^{5} erg cm^{-2} sec^{-1}
W cm^{-2}	10^{7} erg cm^{-2} sec^{-1}

1.2 RADIATION LAWS

1.2.1 Planck's Law

The German physicist Max Planck (1858–1947) showed, in 1900, that the spectral energy density emitted by a blackbody at temperature T is given by the Planck function according to the following relations, the first being expressed on the basis of frequency ν and the second on the basis of wavelength λ of the radiation:

$$B_\nu(T) \;=\; \frac{2h\nu^3}{c^2(e^{h\nu/kT} - 1)} \tag{1.14}$$

$$B_\lambda(T) \;=\; \frac{C_1\lambda^{-5}}{(e^{C_2/\lambda T} - 1)} \tag{1.15}$$

The values of the constants are (Condon and Odishaw, 1967)

h = Planck's constant = 6.6256×10^{-27} erg sec
k = Boltzmann's constant = 1.3805×10^{-16} erg deg^{-1}
c = speed of light $in\ vacuo$ = 2.998×10^{10} cm sec^{-1}
C_1 = first radiation constant = $2hc^2$ = 3.74150×10^{-5} erg cm^2 sec^{-1}
C_2 = second radiation constant = hc/k = 1.43879 cm deg

Curves of $B(T)$ versus wavelength for a number of temperatures are shown in Fig. 1.4. Extensive tabulations of the Planck function for many values of temperature have been issued by various authors. One of the most comprehensive is that of Pivovonsky and Nagel (1961). Walker (1962) has produced a tabulation of the Planck function on the basis of wave number for various temperatures ranging from 77° to 30,000°K.

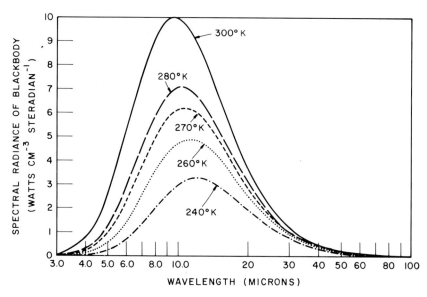

Fig. 1.4 Spectral distribution of radiation emitted by a blackbody at various temperatures.

Two asymptotic forms of the Planck distribution are the Rayleigh–Jeans distribution ($\lambda \to \infty$) and the Wein distribution ($\lambda \to 0$). The behavior of $B_\lambda(T)$ is quite different near the two extremes, and the developments have considerable historical interest. However, the availability of extensive tabulations of the Planck function in its complete form and the possibility of rapid computations by electronic computers make the asymptotic forms of little practical value in the present context.

1.2.2 Kirchhoff's Law

Kirchhoff's law, most clearly enunciated by the German physicist Gustav Kirchhoff (1824–1887) in 1859, states that for a medium in thermodynamic equilibrium, the ratio between the mass emission coefficient and mass absorption coefficient has a value which is independent of the nature of the material and is dependent only on wavelength of the radiation and temperature of the medium. At each point we have

$$j_\lambda/\kappa_\lambda = B_\lambda(T) \tag{1.16}$$

where $B_\lambda(T)$ is the Planck function given Eq. (1.15).

Although the above discussion is based on the concept of a local region of a medium, such as a given volume of air which is in thermodynamic equilibrium, Kirchhoff's law is also applicable to an element of surface which is in thermodynamic equilibrium. For this case, we have

$$J_\lambda/A_\lambda = B_\lambda(T) \tag{1.17}$$

where A_λ is the absorptivity of the surface and J_λ is the areal emission coefficient.* Since the monochromatic emissivity ϵ_λ is defined by the relation

$$J_\lambda = \epsilon_\lambda B_\lambda (T) \tag{1.18}$$

we have for the surface the absorptivity equal to the emissivity. Thus

$$A_\lambda = \epsilon_\lambda \tag{1.19}$$

For a blackbody surface

$$A_\lambda = \epsilon_\lambda = 1 \tag{1.20}$$

at all wavelengths. A "graybody" surface is characterized by the relation

$$A_\lambda = \epsilon_\lambda < 1 \tag{1.21}$$

---

* This term is chosen over the more frequently used "emissive power" in order to be consistent with the discussion leading to Eq. (1.16). The physical quantity involved is the energy emitted by the surface, the units being energy per unit area, unit time and unit wavelength interval (e.g., cal cm^{-2} sec^{-1} μm^{-1}).

A note on the limits of applicability of Kirchhoff's law to the terrestrial radiation field of the atmosphere is appropriate. The law requires thermodynamic equilibrium, one characteristic of which is that the radiation field be isotropic. Obviously, the radiation field for the atmosphere as a whole is not isotropic. On the other hand, the field in a localized volume of the troposphere or stratosphere is approximately isotropic, and it is in the context of this local thermodynamic equilibrium that Kirchhoff's law is applicable to the atmosphere. A second characteristic of local thermodynamic equilibrium is that the populations of atomic and molecular states be those of their equilibrium distribution. In such a case, the energy transitions are controlled by molecular collisions, and not by interactions of particles with the radiation field itself. In the atmosphere, molecular collisions dominate the energy transitions at all altitudes below 60–70 km, indicating that local thermodynamic equilibrium is a good approximation through more than 99% of the mass of the atmosphere.

1.2.3   Stefan–Boltzmann Law

Two other important relations can be derived from Planck's law for blackbody radiation. By integrating the monochromatic blackbody function of Eq. (1.15) over the entire wavelength range from 0 to ∞ we obtain an expression for the total rate of energy emission E by a unit area of blackbody surface in terms of its absolute temperature T as

$$E = \sigma T^4 \tag{1.22}$$

Here σ is the Stefan–Boltzmann constant, a generally accepted value* of which is 5.6697×10^{-12} W cm^{-2} deg^{-4}. This important relation was first obtained from experiments by the Austrian physicist Josef Stefan (1853–1893) in 1879 and was derived 5 years later from thermodynamic theory by another Austrian physicist Ludwig Boltzmann (1844–1906).

A corollary of the Stefan–Boltzmann law is that the rate of total energy emission for a unit area of a graybody surface is given by

$$E = \epsilon \sigma T^4 \tag{1.23}$$

where $\epsilon < 1$ is the (wavelength-independent) emissivity of the graybody surface.

---

* One note of caution on the accuracy of the Stefan–Boltzmann constant is in order. There has been generally a discrepancy of the order of 1% between the theoretically determined value given here and that determined from experiment, the latter being the higher. However, a recent measurement by Blevin and Brown (1971), at the Australian National Standards Laboratory, shows consistency with theory within their experimental uncertainty of about 0.2%.

1.2.4 Wien Displacement Law

A second important relation is obtained by differentiating the Planck function with respect to wavelength, setting the result equal to 0, and thereby determining the value λ_{max} at which $B_\lambda(T)$ is a maximum. This yields Wien's displacement law, first derived by the German physicist Wilhelm Wien (1864–1928) in 1894, as

$$\lambda_{max}T = \text{const} \tag{1.24}$$

The observed value of the constant is 0.2897 cm deg if λ is in centimeters.

The dependence of the position of the maximum of the blackbody function on temperature, as given by the Wien displacement law, can be seen by the blackbody curves of Fig. 1.4.

1.3 MILESTONES IN RADIATION RESEARCH

Perhaps the earliest investigation of the Sun itself was that of Galileo Galilei (1611) following the invention of the telescope. Galileo, his contemporaries, and his successors up to about the beginning of the nineteenth century made extensive visual observations of the Sun in white light, studying in detail the behavior of sunspots and observing the rotational characteristics of the Sun. In spite of two centuries of visual observations, however, it was not until the middle of the 1800s that the eleven-year sunspot cycle was detected and its period accurately determined. In 1666 Sir Isaac Newton discovered the spectral character of sunlight, and decomposed, by the use of glass prisms, the visible spectrum into its monochromatic components. He suggested the corpuscular theory of the nature of light, which by obvious means would explain mirror-type (specular) reflection and the propagation of light in a straight line. He believed that the elementary wave theory, which had been proposed by the Dutch physicist, mathematician, and astronomer Christian Huygens (1629–1695), could not explain the rectilinear propagation of light and that a wave required some transmitting medium which was not evident in his observations. The electromagnetic character of light was not to be discovered for another two centuries. The idea of wave interference and knowledge of the very short wavelengths of light were also nonexistent in Newton's day.

Other notable discoveries during the very prolific seventeenth century were the law of refraction of light by the Dutch professor of mathematics Willebord Snell (1591–1626) (Snell's law, 1621) and the phenomenon of diffraction, discovered independently by the Jesuit professor of mathematics at Bologna, Italy, Francesco Grimaldi (1618–1683), and by Robert Hooke (1635–1703), a physicist of the Stuart School of Scientists in England and Curator of the Royal Society. It is interesting that Robert Hooke claimed discovery of the inverse square law of gravitation before its discovery by Hooke's contemporary and fellow member of the Stuart School of Scientists, Sir Isaac Newton. Other members of the Stuart school at the time were the famous architect Sir Christopher Wren (1632–1723) and Sir Edmund Halley (1656–1742), discoverer of the comet known by his name. Also in the seventeenth century, the Danish astronomer Olaus Roemer (1644–1710), by observations of the apparent change of period of revolution of the moons of Jupiter as the Earth was moving toward and away from Jupiter, showed that light has a finite speed and obtained an approximate value of its magnitude (1676). Huygens, the founder of the wave theory of light, discovered and explained the phenomenon of double refraction in crystals (1678), and discovered but did not explain the polarization of light. Newton's observations of colors in thin films, such as soap bubbles, led to his observing the phenomenon we now know as Newton rings, but their explanation required the wave theory to which he did not subscribe.

The eighteenth century was phenomenally devoid of important discoveries on either the properties of the Sun or the nature of light. Perhaps the most important optical discovery of that century was the observation by the English astronomer James Bradley (1693–1762) that the addition of the vector velocities of the Earth and the light from a star may cause an apparent shift of the position of the star (stellar aberration). The list of great discoveries which filled the 19th century was led by the work of the London physician Thomas Young (1773–1829), from 1800 to 1803, in proposing the principle of interference of waves, by which he was able to completely explain Newton's rings and attempt to explain the problem of diffraction. Young's idea of transverse waves was transmitted by D. F. J. Arago (1786–1853) to a civil engineer, Augustin Fresnel (1788–1827), at the Academy of Sciences in Paris. In 1818, in an entry to a contest for the best essay on optical diffraction, Fresnel showed that all the then known phenomena of optics could be explained on the basis of transverse wave vibrations. Only by the use of transverse waves was it possible to explain the phenomenon of the polarization of light, which had already been rediscovered (after Huygens), in 1808, by Etienne Malus (1775–1812) at the Ecole Polytechnique in Paris. Fresnel, undoubtedly the greatest optical investigator of the period, went on to outline the main features of physical optics essentially as we know them today. In addition to his explanation of diffraction, his best-known contribution is probably the well-known Fresnel laws of reflection and refraction. One year after Malus' observation of the polarization of light by reflection, Arago, also at the Ecole Polytechnique in Paris, discovered that the scattered light from the sunlit sky is partially polarized. The polarization of skylight was studied in some detail during the few decades following Arago's discovery. Arago himself discovered, also in 1809, that at one point in the sky the light is completely neutral, i.e., the polarization vanishes, a point which is now known as the Arago point, and that the maximum polarization occurs at a point 90° from the Sun in the vertical plane through the Sun. These points will be discussed further in Chapter 7. A second neutral point, the Babinet point, was discovered by the French meteorologist and physicist Jacques Babinet (1794–1872) in 1840, a full three decades after the first observation of the Arago point. The Babinet and Arago points are normally located about 20° above the Sun and the antisolar point, respectively, and are relatively easy to observe with a Savart polariscope. The famous Scottish physicist, Sir David Brewster (1781–1868), on being informed of Babinet's discovery, was led by considerations of symmetry to predict the existence of a neutral point below the Sun. Because of the very bright sky and weak polarization in that region of the sky, however, visual observation in the Savart polariscope is very difficult. In searching for the predicted neutral point Brewster was "perplexed beyond measure with the feeble and uncertain indications of the polariscope." However, by setting the polariscope inside a long dark passageway and observing the sky beneath the Sun from the end of the passageway, Brewster obtained a distinct view of the neutral point on 28 February 1842. After many attempts, Babinet was able to confirm its existence on 23 July 1846.

Of the other radiation work during the first half of the nineteenth century, most important in the present context is Joseph von Fraunhofer's (1787–1826) publications, in 1814, of his extremely careful and detailed spectroscopic observations of the solar spectrum showing 574 dark spectral lines. The explanation of the Fraunhofer lines, as being due to absorption in the atmosphere of the Sun, was given by G. R. Kirchhoff some 45 years later.

Knowledge of the solar radiation regime advanced very rapidly during the last half of the nineteenth century, due to work on the three fronts of instrumentation, observation, and theory. In 1852 Sir George Stokes published a new theoretical representation of light in terms of four parameters, all of which have the dimensions of intensity and are, therefore, additive for mixtures of different streams of light. It is remarkable that Stokes' representation was largely unused for almost 100 years, after which time our own contemporary mathematical physicist and astronomer, S. Chandrasekhar, resurrected the concepts and developed, thereby, a powerful theoretical tool for the study of radiation in stellar and planetary atmospheres. The method introduced by Chandrasekhar, and the results which have been obtained by several authors through use of the Stokes parameters, will be discussed at length during the course of this book. The essential electromagnetic theory of radiation, as we know it today for macroscopic systems, was developed by the great English theoretical physicist James Clerk Maxwell (1831–1879) about 1860.

The theory of radiative transfer in a scattering medium was put on a firm theoretical basis by another prominent English physicist John William Strutt (1842–1919), later Lord Rayleigh, in 1871 through his famous explanation of the polarization and color of the light from the sunlit sky. Since Rayleigh's theory is postulated on the assumption that the scattering particles are of small dimensions compared to the wavelength of the radiation, such small particles as molecules and very small aerosol particles have become known as Rayleigh particles, and an atmosphere composed of such small particles is termed a Rayleigh atmosphere.

Although Rayleigh's theory explained many of the observed features of skylight, it did not predict the existence of the neutral points as they had already been observed by Arago, Babinet, and Brewster. The French physicist J. L. Soret attempted, in 1888, to explain the observed neutral points as being due to secondary scattering of radiation in the atmosphere, Rayleigh's model having considered only primary (single) scattering by the gaseous molecules. Even with a greatly simplified model, Soret did show that the neutral points could be caused by secondary scattering. Later attempts to account for secondary and higher-order scattering met with only limited success until the problem was solved exactly for quite realistic atmospheric models, by Chandrasekhar, about 1950.

The optical effects produced by the volcanic ash injected into the atmosphere by the eruption of the volcano Krakatao, in 1883, generated a flurry of interest in skylight measurements. The neutral points, particularly the Babinet and Arago points, were observed extensively for several years following the eruption, and they were found to be shifted from their normal positions by several degrees in the direction away from the Sun and antisolar point, respectively. Additional neutral points not normally present were observed to be positioned horizontally on either side of the Sun and antisolar point, and the magnitude of the maximum polarization was greatly reduced by the volcanic dust. During the years after the Krakatao eruption, the French experimental physicist Marie Alfred Cornu (1841–1902) brought the method of measuring polarization visually to a high degree of perfection by his photopolarimeter (1890) based on a combination of independently rotatable Wollaston and Nicol polarizing prisms. The same system, with a slight improvement by Martens introduced in 1900, is often used for visual observations today. By atmospheric observations with his photopolarimeter, Cornu first observed the now well-known fact that the degree of polarization of skylight varies with the wavelength of radiation.

Details of the solar spectrum were investigated in the pioneering work of Samuel Pierpont Langley (1834–1906) of the Smithsonian Institution, the same Langley to whom we owe our system of the time zones and for whom, in the United States, the unit of radiation (cal cm^{-2}) is named. As early as 1873 Langley became interested in the Sun, and spent the next three decades studying its radiation. In 1880 he invented the extremely sensitive and versatile bolometer for measuring solar spectra. Just at the turn of the century (1900, 1902), he published some remarkable records of the solar spectrum through the wavelength range of 0.3 to 5.3 μm.

The solar spectrum was also studied by the American physicist H. A. Rowland (1848–1901) and by the French physicist A. H. Becquerel (1852–1908). Rowland produced, about 1897, a monumental map of the spectrum over the wavelength range of 0.2975 to 0.7331 μm. By the use of an infrared-sensitive phosphor, Becquerel discovered, in 1883, that the incident solar radiation is partially absorbed by water vapor in the Earth's atmosphere.

Radiation instrumentation was advanced significantly during the waning years of the nineteenth century by the invention by Knut Ångström (1857–1910), second member of the distinguished Swedish family of physical scientists, of the Ångström electrical compensation pyrheliometer (1899). The Ångström pyrheliometer is still used as the standard for absolute radiant energy determinations in many countries of the world. The principle was later employed by the third member of the family, Anders K. Ångström (1888–) to construct a new type of instrument (pyrgeometer) for measuring the nocturnal longwave atmospheric radiation.

Because of the extremely rapid and extensive amplification of knowledge of radiation and its measurement during the twentieth century, it is impossible to summarize the advances in a short discussion. However, from a solar-radiation-measurement standpoint, the period can be conveniently divided at about the time of World War II (1939–1945). The period before that time saw the development of many different kinds of radiation instruments, some of which have been rendered obsolete, but others remain as either standard instruments against which operational instruments are calibrated or as operational instruments themselves. In the standard category are the water-flow pyrheliometer (now in disuse) and the silver-disk pyrheliometer. (The Ångström electrical compensation pyrheliometer, also a primary reference, preceded this period by a few years.) Operational pyrheliometers are mainly of the Eppley, Linke–Feussner, and Savinov–

Yanishevsky types, but the Michelson pyrheliometer is used relatively extensively in the U.S.S.R. Although later modifications have been introduced, the basic designs of these were developed before 1940.

Of the numerous pre-World War II designs of pyranometers, the principal ones used in radiation networks are those of Kimball–Hobbs, which is the basis of the several Eppley 180° pyrheliometers (now discontinued), Moll–Gorczynski, the basis of the Kipp and Zonen instrument, and the Robitzsch and Bellani pyranometers (CSAGI, 1958). The Yanishevsky pyranometer, the main instrument of the operational network in the U.S.S.R., is a postwar development (Kondratyev, 1969).

In contrast to many types of other physical instruments, those for measuring solar radiation were not advanced much during World War II. Since 1945, however, larger numbers of research scientists working with the increased budgets generated by a combination of international tensions and space explorations have produced a new order in radiation investigations. Military programs have sponsored extensive research work on radiation sensors, transmitting or absorbing materials, and methods of data handling, as well as on radiative transfer theory. Unfortunately, many of those results have not found their way into the open literature. However, the work associated with the United States space program and with the development of meteorological satellites has produced tremendous advances in instrumentation and techniques of measurement, as well as in theoretical analysis of radiative transfer.

Radiation instrumentation at a more purely operational level has also been advanced in the postwar period. The incorporation of temperature compensation into the circuits of the Eppley instruments, both pyrheliometers and pyranometers, has been a major improvement. New and improved methods of calibrating pyranometers have been instituted, and the first steps toward programs for maintaining the WMO solar radiation networks are being implemented. New types of pyranometers have appeared: the Star pyranometer, the Yanishevsky pyranometer, the Eppley models, and others. Multichannel spectral (filter) radiometers have been developed for precise solar radiation measurements at the ground, in aircraft, and on space vehicles. An operational ultraviolet pyranometer for surface measurements of ultraviolet flux was introduced a few years ago. There have also been considerable advances in the routine measurement of the components of terrestrial radiation. Several improvements in operational instruments in the Soviet Union have recently received attention, the main ones being the use of thermopiles instead of simple resistance strips in the Ångström-type pyrheliometer, a redesign of the collimator tube for pyrheliometers, sealing the enclosed air space of the pyranometer, and the installation of black disks at the level of the pyranometer receiver for protection from internal reflections. Experimentation toward the development of a new distillation-type instrument has been started in Finland, as has the development in Belgium and the United States of new standard instruments for irradiance measurements. The use of a retardation plate-polarizer combination was instituted for measurements of skylight polarization, and the measurement of polarization was facilitated by the use of a simple rotating analyzer as the basis of the optical system.

Perhaps even greater advances have been made in the methods of recording and processing data. New technology based on electronic computers has rendered obsolete many of the time-consuming data-handling methods extant before the last World War. Similarly, the newer digital-type recording equipment points the direction toward which radiation data acquisition methods must move, and the costs of such systems are being reduced to render them applicable to many of the radiation measurements of an operational nature.

1.3.1 Highlights of Solar Radiation Research

A list of important events which have occurred in solar radiation research since the early nineteenth century is given below. It is realized that the list is not complete (some additional ones are given by Drummond, 1970), and the choice of events is somewhat subjective.

1825	Herschel's pyrheliometer (actinometer) developed: first instrument in which the rate of cooling was introduced into solar radiation measurements
1837	Invention of Pouillet's pyrheliometer: first use of the term "pyrheliometer", solar constant measurement by Pouillet yielded 1.76 cal cm^{-2} min^{-1}
1838 to 1840	Invention of first photographic sunshine recorders by T. B. Jordan
1840	Sir John Herschel obtained the first photograph on a glass plate (a picture of a telescope)
1845	First daguerrotype of the Sun was taken by Foucault and Fizeau
1859	Development of Kirchhoff's law
1879	Experimental development of Stefan (later called Stefan–Boltzmann) law. Device introduced by Stokes for use of cards with the Campbell sunshine recorder; instrument called Campbell–Stokes sunshine recorder since that time
1881	Observations from top of Mt. Whitney extended known solar spectrum to 1.8 μm. Value of solar constant estimated to be greater than 3 cal cm^{-2} min^{-1}
1883	Becquerel obtained map of solar spectrum out to 1.4 μm
1884	Boltzmann's derivation of Stefan–Boltzmann law on basis of thermodynamic theory
1885	Invention of photographic sunshine recorder by J. B. Jordan. Invention of new form of polarimeter and of pole-star recorder, both by E. C. Pickering

1886	Abney published table of 429 lines of solar spectrum at wavelengths below 1 μm. Use of first thermoelectric pyrheliograph by Crova. Invention of new type of radiometer by Knut Ångström
1891	Concept of Maring–Marvin sunshine recorder by D. T. Maring
1893	Invention of electric compensation pyrheliometer by K. Ångström; improved versions described in 1896 and 1899
1894	Development of Wein displacement law
1897 to 1898	Langley's determination of location and approximate strength of over 700 lines of solar spectrum between 0.4 and 6 μm
1898	Invention of Callendar pyranometer; improved version brought out in 1905
1900	Development of Planck's law. Measurements of total solar radiation begun in Washington, D.C., by Smithsonian Institution
1902	Monochromatic observations begun by Langley in Washington, D.C. Estimates of solar constant varied between 1.75 to 4 cal cm^{-2} min^{-1} Work started on a mercury pyrheliometer, which was later developed into the secondary standard silver-disk pyrheliometer of Abbot
1903	Smithsonian Institution started construction of the water-flow pyrheliometer, which was perfected into the primary standard about 1910
1904	First attempt, by Langley, to correlate weather with solar radiation
1905	Einstein developed quantum theory of radiation. Ångström electrical compensation pyrheliometer adopted as a standard at Meteorological Conference, Innsbruck, and by Solar Physics Union, Oxford. Smithsonian solar observation expedition to Mt. Wilson. Additional expeditions in 1906, 1907, 1908
1906	Best estimate of solar constant: 2.1 cal cm^{-2} min^{-1}. Measurements of reflectance of stratus cloud near Mt. Wilson gave value of 65%
1907	Weather Bureau solar radiation observatory established in Mt. Weather, Virginia
1908	Development of Mie theory of scattering. Invention of Michelson pyrheliometer. Start of study of transmission of water vapor by F. E. Fowle
1909	Abbot's first expedition to Mt. Whitney. First construction of Abbot silver-disk pyrheliometer (modified to long tube in 1927). Summary of observations from 1905 to 1909 gave value 1.924 cal cm^{-2} min^{-1} for solar constant
1910	Invention of Marvin pyrheliometer. Value of solar constant estimated as 1.922 cal cm^{-2} min^{-1}. Pyrheliometric observations started on Mt. Fuji, Japan, and at Madison, Wisconsin
1911	Solar observation station established at Bassour, Algeria. Solar radiation station established at Lincoln, Nebraska
1912	Radiation measurements in Algeria showed 20% reduction of sunlight due to volcanic ash from Mt. Katmai eruption in Alaska. Experiments started with "water-stir" pyrheliometer. Observation station established at Harvard College Observatory, Arequipa, Peru
1913	Mean value of solar constant given as 1.933 cal cm^{-2} min^{-1}. First revision of Smithsonian pyrheliometric scale, based on the water-flow pyrheliometer. Abbot constructed hemispherical pyrheliometer, and a recording balloon pyrheliometer which was carried to 45,000 feet

altitude. First daylight illumination measurements made at Mt. Weather, Virginia with Sharp–Millar photometer

1914   Balloon flights yielded solar constant of 1.88 cal cm^{-2} min^{-1}

1915   First vacuum bolometer constructed

1916   Pyranometer for measuring global radiation devised. Abbot's "solar cooker," for using direct solar radiation for cooking food, installed.

1917   Ultrasensitive vacuum bolometer constructed. Observation station installed on Hump Mountain, North Carolina

1918   Solar observation station installed near Calama, Chile. New Instrument constructed for comparing brightness of Sun and sky. Balloon measurement, by Aldrich, of cloud albedo at Arcadia, California, gave value of 78%. "Short method" of solar constant determination devised and put into use

1919   Honeycomb (melikeron) pyranometer constructed. Marvin pyrheliometer and A. K. Ångström's electrical compensation pyrgeometer invented

1920   Calama, Chile, station moved to Mt. Montezuma, Chile, where it was operated continuously until 1955. Mt. Wilson station moved to Mt. Harqua Hala, Arizona; abandoned in 1925: mean of 1244 observations at various locations from 1912 to 1920 yielded a solar constant of 1.946 cal cm^{-2} min^{-1}

1922   Dorno's pyrheliograph invented

1923   Invention of Kimball–Hobbs pyranometer (forerunner of first Eppley pyranometers). Invention of Moll thermopile. First weather forecasts based on (indicated) variations of solar constant made by H. H. Clayton

1924   Moll thermopile used by Gorczynski for first Moll–Gorczynski pyranometer, later known as Kipp solarimeter

1925   Solar observation station on Table Mountain, California, established (discontinued in 1958). Strong questioning of Smithsonian Institution's determinations of solar variability begun

1926   Solar observation station established at Mt. Brukkaros, South West Africa; abandoned in 1931

1927   Double water-flow pyrheliometer developed by Shulgin

1929   Method introduced by Kalitin for measuring radiative flux from various zones of the sky. Richardson developed method of measuring reflectivity of natural surfaces from aircraft

1931   Method developed by Kalitin for measuring albedo of natural surfaces from ground

1932   Double water-flow pyrheliometer was adopted as standard by Smithsonian Institution, and first revision of 1913 pyrheliometric scale. (Although the revised scale was reconfirmed in 1934, 1947, and 1952, and showed the 1913 scale to be approximately 2.3% too high, it was never generally adopted.) First standard design of Robitzsch pyranometer developed

1933   Solar observation station established on Mt. St. Katherine on Sinai Peninsula; abandoned in 1937

1939   Solar observation station established on Burro Mountain, New Mexico; abandoned in 1946

1945	Solar observation station established at Camp Lee, Virginia; abandoned in 1947
1948	Menzel developed sensitive sky photometer
1950	Observation begun at Table Mountain with Menzel sky photometer
1952	The silver-disk pyrheliometer recommended as an instrument for measuring direct solar radiation by Subcommission on Actinometry, World Meteorological Organization, Brussels. Best estimate of solar constant: 1.94 cal cm^{-2} min^{-1}
1953	Invention of sunshine switch by Foster and Foskett of U.S. Weather Bureau
1954	Weather Bureau method of calibrating Eppley pyranometers changed. Old method utilized natural solar radiation; new method uses artificial radiation in an integrating sphere (similar methods later introduced by Canadian Meteorological Service and Eppley Laboratory). Reevaluation of solar constant, by Johnson, through use of new rocket observations in ultraviolet and revised estimates in infrared. Best estimate given as 2.00 cal cm^{-2} min^{-1} with a probable error of 2%
1956	Statistical study of 30 years of solar radiation data of Mt. Montezuma and Table Mountain by Sterne and Dieter showed root mean square of real changes of the solar constant no greater than 0.0032 cal cm^{-2} min^{-1}. "International Pyrheliometric Scale 1956" recommended by International Radiation Conference, Davos
1957	"International Pyrheliometric Scale 1956" put into effect; original Ångström scale increased by 1.5% and 1913 Smithsonian scale decreased by 2.0%
1961 to 1968	Balloon program for solar constant measurements conducted in Soviet Union
1962	Adoption of the Campbell–Stokes sunshine recorder as an "Interim Reference Sunshine Recorder" by the Commission on Instruments and Methods of Observation, W.M.O. Evaluation of solar radiation pressure on motion of satellite, by Jacchia and Slowey, provides value of solar constant as 2.00 cal cm^{-2} min^{-1}
1965	Introduction of Eppley precision pyranometer. Development of automatic control of Ångström pyrheliometer by Marsh
1966 to 1969	Program using aircraft, balloons, and spacecraft for direct measurement of solar constant and its spectral distribution conducted in the United States
1968	Value of solar constant as 1.952 cal cm^{-2} min^{-1} obtained by Laue and Drummond from rocket aircraft flight to 83 km altitude on 17 October 1967
1969	Introduction of Eppley black and white (star-type) pyranometer. Previous model, based on Kimball–Hobbs design, discontinued
1971	Announcement by Thekaekara and Drummond that 1.940 cal cm^{-2} min^{-1} adopted as an engineering design value of the solar constant by the U.S. National Aeronautics and Space Administration

REFERENCES

Ambartsumian, V. A. (ed.) (1958). "Theoretical Astrophysics" (Transl. from Russian by J. B. Sykes). Pergamon, New York.

Blevin, W. R., and Brown, W. J. (1971). A precise measurement of the Stefan–Boltzmann constant. *Meterolog.* 7, 15–29.

Chandrasekhar, S. (1950). "Radiative Transfer." Oxford Univ. Press (Clarendon), London and New York.

Condon, E. V., and Odishaw, H. (1967). "Handbook of Physics." McGraw–Hill, New York.

CSAGI (1958). Radiation instruments and measurements. Part IV, "IGY Instruction Manual," pp. 371–466. Pergamon, Oxford.

Drummond, A. J. (1970). A survey of the important developments in thermal radiometry. *In* "Advances in Geophysics," (A. J. Drummond, ed.), Vol. 14. Academic Press, New York.

Goody,R. M. (1964). "Atmospheric Radiation, I, Theoretical Basis." Oxford Univ. Press (Clarendon), London and New York.

Kondratyev, K. Ya. (1969). "Radiation in the Atmosphere." Academic Press, New York.

Pivovonsky, M., and Nagel, M. R. (1961). "Tables of Blackbody Radiation Function." Macmillan, New York.

Walker, R. G. (1962). Tables of the blackbody radiation function for wavenumber calculations. Res. Rep. AFCRL-62-877, U.S. Air Force Cambridge Res. Lab., Hanscom Field, Massachusetts.

World Meteorological Organization (1965). Guide to meteorological instruments and observing practices. WMO No. 8 TP 3, Geneva, Switzerland.

CHAPTER TWO

Radiation Sensors and Sources

Radiation sources are necessary for the testing and calibration of radiation measuring instruments. Likewise the usefulness of an instrument for a particular measurement is largely determined by the detector on which the instrument is based. The technology, particularly for radiation detectors, developed very rapidly during World War II because of the need to detect the existence and location of personnel and other military targets, and the military has continued to sponsor a great deal of activity in infrared technology (Wolfe, 1965). In fact, although little information is available on the point, it is known that some of the results are reported only in the classified literature. The space program and projects in remote sensing of the environment have been fertile spawning grounds for improvements in radiation technology, and significant resources have been applied by industrial concerns in the field.

It is impossible to cover all aspects of radiation sensors and sources in a short space but good surveys, particularly for the infrared region, are available (e.g., Kruse et al., 1962; Bolz and Tuve, 1973; Brown, 1965; Jamieson et al., 1963; Smith et al., 1968). Consequently, this discussion will be restricted to those aspects which are most applicable to instruments in atmospheric radiation.

2.1 RADIATION SENSORS

Radiation sensors or detectors of most use in atmospheric problems may be broadly classified as thermal detectors and photoelectric detectors of various types. The two types will be discussed separately.

2.1.1 Thermal Detectors

The transfer of radiant energy into heat energy, with a consequent rise of temperature of some material, is the mode of operation of thermal detectors. They respond only to total energy absorbed, and are thus, at least theoretically, nonselective as to the spectral distribution of the energy. Because of the limitations of absorbing materials, this nonselective feature is difficult to achieve completely in operation, but it is more closely achieved in thermal detectors than in any other type.

The main types of thermal detectors are calorimeters, thermocouples or thermopiles, and bolometers.

Calorimeters A direct determination of the amount of radiant energy absorbed by measurements of a temperature change in a material is the basis of a calorimeter detector. As will be seen below, the water-flow and silver-disk pyrheliometers of the Smithsonian Institution and the Marvin pyrheliometer, utilize this principle. Variations of this basic type are the Robitzsch pyranometer and Michelson pyrheliometer, in which the change of temperature causes a distortion of a bimetallic element, the Ångström electrical compensation pyrheliometer, in which the change of temperature due to absorption of radiation by one element is reproduced by a measured amount of electrical energy in a companion element, and the Golay cell, in which the rise of temperature of an enclosed gas is observed as a change of pressure inside the cell. The main advantage of a calorimetric-type detector is its basic simplicity, while it suffers from being relatively insensitive and having a slow response.

Thermocouples and Thermopiles As discovered by Peltier in 1834, a difference in temperature between a junction of two dissimilar metals and a reference junction (a thermocouple) produces an electromotive force (emf) across the junctions. The amount of emf depends on the types of metals, but it is so small for a single thermocouple as to be difficult to measure accurately. The voltage may be built up, however, by connecting a number of thermocouples in series, thereby making a thermopile. The convenience of an easily measurable voltage output, coupled with the approximate nonselectivity of a thermal detector, has made the thermopile the most frequently used detector in atmospheric radiometers. It is the basic sensor of the Eppley, Linke–Feussner, and Savinov–Yanishevsky pyrheliometers, and the Eppley, Star, Moll–Gorczynski, and Yanishevsky pyranometers, to mention a few.

Bolometers The bolometer, first developed by S. P. Langley in 1879–1880, is one of the most sensitive of radiation detectors of the nonselective type. Langley (1900) claimed that it could register a temperature change of

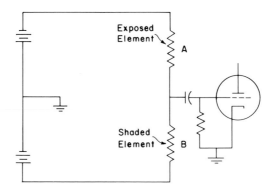

Fig. 2.1   The electrical connections of a bolometer.

0.00000001°C, although, as far as is known, this has not been confirmed. Its operation is based on the change of resistance of a metal or semiconductor with temperature. The normal configuration of a bolometer circuit for radiation measurements is shown in Fig. 2.1. The two resistance elements A and B constitute two arms of a Wheatstone bridge. If element A is exposed to radiation while element B is shaded, the relative temperature difference which is generated between them results in an imbalance of the bridge. By proper calibration, the amount of imbalance may be interpreted in terms of the flux of radiant energy incident on the exposed element. The theory of the bolometer has been discussed recently by Strong and Lawrence (1968) and by Kruse *et al.* (1962).

Bolometers are of three general types: metal, semiconductor, and superconducting. The last of these is not well adaptable to atmospheric radiation measurements, as it requires temperatures of near absolute zero. The change of resistance with temperature is much higher for semiconductor materials (thermistors) than for metals, a fact which makes thermistor bolometers more frequently used in atmospheric measurements than those of metal. In order to achieve fast response, the thermistors are generally made in the form of flakes of about 10 μm thickness, and are mounted on a heat-dissipating substrate. The flakes are normally made of sintered oxides of manganese, cobalt, and nickel.

The high sensitivity and lack of spectral selectivity of the bolometer have made it particularly useful in spectral measurements of atmospheric radiation. Langely himself used his bolometer in the early 1900's to establish the spectral distribution of energy from the Sun, and it was a basic component of the instrumentation used by C. G. Abbot and his colleagues of the Smithsonian Institution to measure the solar constant (see Chapter

9). The relatively fast response which can be achieved is exploited in thermal mapping for remote sensing applications.

Pyroelectric Detectors   This is a recent development in detector technology, but it appears to have interesting possibilities for infrared atmospheric measurements. Radiation absorbed by the pyroelectric crystal is converted to heat, thereby altering the lattice spacings within the crystal and causing a change of the spontaneous electric polarization of the crystal. If electrodes on the surfaces of the crystal are connected through an external circuit, the current generated is proportional to the rate of change of temperature in the crystal. Thus a modulation of the incident radiation is necessary for operation of the device.

The most attractive features of the pyroelectric detector are its extremely wide spectral sensitivity and its fast response. Although it is a thermal device, the sensitivity is controlled mainly by window materials or other optics of the system. Useful sensitivity can be made to cover the entire range of 0.2 to 1000 μm. The rapid response (typical rise time 0.5 to 5 nsec) makes the device particularly useful for infrared laser application. However, the system response time of at least one pyroelectric radiometer on the market* varies from 0.1 to 10 sec for 90% response as the sensitivity range is varied from 100 mW cm^{-2} to 0.1 μW cm^{-2}. Presumably the degradation of response time is due to the system electronics and not to the pyroelectric crystal itself.

2.1.2   Photodetectors

The great advantage of photodetectors is that the sensor is activated by discrete events of photons striking the material, and not by a change of temperature because of absorption of the radiation, as in thermal detectors. Although not every one of the photons incident on a photo-detector causes the event to occur (the quantum efficiency is typically 0.1%–10%) and the wasted photons simply go to heat in the detector, the amount of heating by the noneffective photons is negligible for most photodetectors. The advantages of sensing discrete events is that much faster responses and higher sensitivities can be achieved than with thermal devices.

The three principal types of photo detectors are photovoltaic, photo-conductive, and photoemissive cells.

Photovoltaic Detectors   These are the simplest of the photodetectors, and they provide an added advantage of yielding measurable voltages without

---

* Molectron Corp., 177 N. Wolfe Road, Sunnyvale, California 94086.

external power supplies when illuminated by visible or near-ultraviolet radiation. The most common of photovoltaic detectors is the selenium cell, which finds frequent application in photographic light meters and illuminometers. It is described in more detail in the chapter dealing with illumination (Chapter 6). Certain other types of materials (e.g. indium antimonide, indium arsenide, gallium arsenide) are sometimes used in the photovoltaic mode for detecting radiation in the 1 to 5 μm region (Bolz and Tuve, 1973). The spectral response of an indium–antimonide detector cooled to 77°K and operated in the photovoltaic mode is shown by curve F in Fig. 2.2. Another photovoltaic device is the silicon solar cell, which finds most application for power generation on spacecraft, but solar cells are also used as the sensors on some pyranometers (see Chapter 4).

Photoconductive Detectors   Photocells for measurements in the infrared spectral region are frequently of the type in which the electrical conductance of the material varies with the flux of incident radiation (i.e., photo-

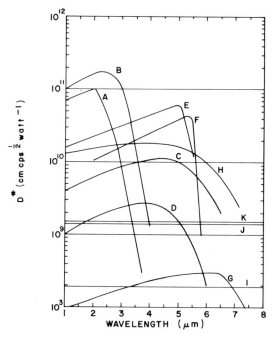

Fig. 2.2   Spectral sensitivity of various types of detectors, expressed in terms of D^* (adapted from Brown, 1965). (A) Lead sulfide: 293°K; (B) lead sulfide: 195°K; (C) lead selenide: 77°K; (D) lead telluride: 77°K; (E) indium antimonide (photoconductive mode): 77°K; (F) indium antimonide (photovoltaic mode): 77°K; (G) indium antimonide (photoelectromagnetic mode): 293°K; (H) gold-doped germanium: 77°K; (I) thermistor bolometer; (J) thermocouple; (K) Golay cell.

conductive type), and a large variety of such cells is available. Unfortunately, many of them require cooling, with the consequent complexity in measurement systems.

The spectral ranges in which the various detectors are useful are shown by the curves of Fig. 2.2 (Brown, 1965), in which the detectivity normalized to unit frequency is plotted as a function of wavelength. The quantity D^* is given by

$$D^* = A^{1/2}/W \tag{2.1}$$

where A is detector area and W is the noise equivalent power of the detector. Some of the detectors are highly wavelength selective, as for instance lead sulfide or indium antimonide, and many require cooling. However, the thermocouple and bolometer, thermal detectors for which curves are shown in the diagram, operate at room temperature and (ideally) show no spectral dependence. The advantage of the solid-state detectors comes, of course, in their much higher sensitivity as shown in the diagram. Typical ranges of the operating parameters of these and some additional photoconductive sensors, as given by Bolz and Tuve (1973), are listed in Table 2.1. The far-infared range is covered advantageously by germanium with various types of impurities (doping materials).

Another type of photoconductive cell which is useful in the visible and near-infrared range is the silicon PIN junction-type photodiode.[†] When operated with no bias voltage this cell acts in the photovoltaic mode, but it is normally used in the photoconductive mode. The spectral response is basically that of silicon, which covers the range from about 0.3 to 1.08 μm with a peak at about 0.9 μm. This cell is very convenient to use, in that it does not require cooling and can be fabricated in many different configurations. Its sensitivity is considerably less than that of photomultiplier tubes, but its response at longer wavelengths and its general operating characteristics make it an attractive choice for certain atmospheric radiation measurements.

Photoemissive Detectors   In both the photovoltaic and photoconductive detectors, the electrons which are dislodged from molecules on impact by photons remain inside the material, and produce a change of voltage across the material or a change of conductance of the material. In photoemissive detectors, however, the electrons are actually ejected from the material. This "photoelectric effect" was first observed by H. Hertz in 1887. Once the electrons are in free space, they can be collected at an anode to give a current flow through the detector, or subjected to high intensity electric

[†] Manufacturer: United Detector Technology, Santa Monica, California.

TABLE 2.1

Typical Values of Operating Parameters of Photoconductive Cells[a]

Detector material	Operating temperature (°K)	Spectral peak (μm)	Useful spectral range (μm)	Detectivity range D^* (cm Hz$^{1/2}$ W^{-1})
Lead sulfide	295	2.4	1.0 to 3.0	0.7 to 1.5 × 10^{11}
Lead sulfide	193	2.7	1.0 to 3.5	2.0 to 7.0 × 10^{11}
Lead sulfide	77	3.2	1.0 to 4.0	0.8 to 2.0 × 10^{11}
Lead selenide	295	3.7	1.0 to 4.5	0.3 to 1.2 × 10^{10}
Lead selenide	193	4.4	1.0 to 5.1	1.5 to 4.0 × 10^{10}
Lead selenide	77	5.0	1.0 to 6.5	1.0 to 3.0 × 10^{10}
Indium antimonide	77	5.3	2.0 to 5.4	2.5 to 5.0 × 10^{10}
Germanium–gold	77	5.0	2.0 to 7.0	3.0 to 6.0 × 10^{9}
Germanium–mercury	28	11.0	2.0 to 13.8	0.7 to 1.5 × 10^{10}
Germanuim–cadmium	21	22.0	2.0 to 23	0.7 to 1.5 × 10^{10}
Germanium–copper	15	24.0	2.0 to 28	0.7 to 1.5 × 10^{10}
Germanium–zinc	12	35.0	2.0 to 38	0.7 to 1.5 × 10^{10}
Mercury–cadmium–telluride	77	12±1	8.0 to 13	2.0 to 6.0 × 10^{9}

[a] As given by Bolz and Tuve (1973).

fields and be accelerated to a second target with enough energy to eject many more electrons. These secondary electrons, in turn, can be accelerated to a third target, with another multiplication of numbers, and so on through as many as 14 or more stages. This process causes a cascade of electrons to build up from a single incident photon, which is responsible for the very high sensitivities obtainable with photomultiplier tubes (more precisely electron multiplying phototubes). Multiplication factors as much as 10^{6} are possible in well-designed photomultiplier tubes.

As Einstein showed in 1905, for an electron to be emitted, the relation

$$h\nu \geq e\phi \qquad (2.2)$$

must be satisfied. Here h is Planck's constant, ν is frequency, e is the unit electrical charge, and ϕ is a constant, the "work function" of the material. Since the atomic binding forces, and therefore the work function, are small in alkali metals (lithium to cesium in the Periodic Table), these substances are all good photoelectric emitters. Even for the best of materials, however, ϕ is sufficiently high to limit the sensitivity of photocathodes to radiation at wavelengths below about 1.0 μm, thereby making photoemissive detectors applicable only to the ultraviolet, visible, and very near-infrared regions of the spectrum.

   The design and construction of photomultiplier tubes has been discussed in a number of books (e.g., Zworkin and Ramberg, 1947; Kruse *et al.*, 1962), and performance specifications of particular tubes are available from the manufactures (e.g., RCA and E. M. I., Varian). Spectral response characteristics of photomultipliers with various types of cathodes, as given in manufacturer specification sheets, are shown in Fig. 2.3. Superimposed on the diagram are lines of quantum efficiency (fraction of incident photons which trigger the emission of an electron). Very high values of 20 to 40% quantum efficiency are attained in the visible and ultraviolet regions, but they are an order of magnitude lower than that in the near-infrared region. The *S*-1 cathode response (curve 11) shows low

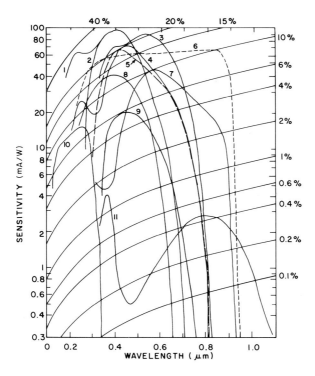

Fig. 2.3 Typical sensitivity as a function of wavelength for various types of photomultiplier tubes (from RCA Tech. Publ. No. PIT-700B, dated 12/71). Curve 1: K–Cs–Sb (133); curve 2: K–Cs–Sb (116); curve 3: Na–K–Cs–Sb (111); curve 4: Na–K–Cs–Sb (138); curve 5: Na–K–Cs–Sb (110: *S*-20); curve 6: GaAs (128); curve 7: Na–K–Cs–Sb (119); curve 8: Cs–Sb (102: *S*-4); curve 9: Ag–Bi–O–Cs (106: *S*-10); curve 10: LiF (125); curve 11: Ag–O–Cs (101: *S*-1).

quantum efficiency and low sensitivity at all wavelengths, but its response extends farther into the infrared than do the others. The S-20 response (curve 5) has been extended into the infrared in one EMI tube sufficiently to yield useful sensitivity at 0.9 μm.

In addition to spectral sensitivity and quantum efficiency, photomultiplier tubes vary in physical size, orientation of the cathode (side-on or end-on), amplification factor, number of stages, amount of dark current, single electron rise time, and other parameters. The choice of a tube for other than the simplest application should be discussed with the vendor. One caution in the use of photomultipliers is that in order to minimize fatigue effects, the current through the cell must be kept very low (a few microamperes at most).

Other photoemissive devices are the vacuum and gas-filled photocells. The simplicity of the vacuum photocell, which consists of a light-sensitive cathode plus collecting anode, makes it a very popular device for many purposes, and its extremely fast response is advantageous in laser applications. However, its sensitivity is too low for most atmospheric radiation measurements. The gas-filled photocell is a more sensitive detector than the vacuum type, but it is not so simple and easy to use. By impressing a high bias voltage across the tube, electrons emitted by the cathode may be accelerated sufficiently to ionize the gas, and thereby increase the gain of the tube by a factor of up to 100 or so. Gas-filled tubes tend to be unstable, however, which decreases their usefulness for accurate radiation measurements, and the response time is much increased by the presence of the gas.

The class of photoemissive detectors includes also television camera tubes of various types, image converters for converting infrared images to visible images, the image orthicon, which is used extensively in satellite instruments, and other similar devices. These tubes are extremely complicated, and a discussion of them is outside the scope of this book.

2.2   RADIATION SOURCES

Interest in sources of radiation is confined here to their use in the calibration of detectors or as sources of illumination. They are conveniently classified purely on the basis of temperature into high-temperature sources, for calibrating solar radiation instruments or providing illumination, and low-temperature sources, for calibrating long wave receivers.

2.2.1   High-Temperature Sources

The most useful sources for calibrating solar radiation instruments are the Sun itself, incandescent lamps, and high-temperature blackbody cavities of various types. Arc lamps are used principally as sources of illumination and for providing line spectra, although high-pressure xenon arc lamps are useful to simulate solar radiation in the visible range of the spectrum. The various types will be discussed below.

The Sun as a Calibration Source   Because of the spectral selectivity of all real radiation receivers and the impossibility of reproducing the solar spectrum by artificial sources, the Sun is the most frequently used source for calibrating solar radiation instruments. International comparisons of working standard instruments are normally made by this method (e.g., Thekaekara *et al.*, 1972). The normal procedure in this case is to set up working standard instruments to be calibrated alongside a standard instrument, for which the calibration constant is known. Then by a ratio of the signals from the two types of instruments taken over several typical periods of operation, the calibration constants of the working standards can be determined.

If the spectral responses of two instruments are the same, then changes of the spectral distribution of the incident radiation, such as those which occur naturally with changing Sun elevation, have no effect in determining the calibration constant of the operational instrument. Calibration of the operational instrument directly from a known energy output from an artificial source, however, does introduce a certain amount of error, even for wide-band receivers such as thermopiles; for phototubes, filtered instruments, and other narrow band receivers it is necessary to take great care in properly accounting for the spectral distribution of the energy from the source.

Incandescent Sources   Standard lamps used for absolute energy determinations are of three different types: (a) standards of spectral radiance, (b) standards of spectral irradiance, and (c) standards of total irradiance. The primary standards of spectral and total radiation in the United States are a group of blackbody sources maintained by the National Bureau of Standards. Secondary standards which have been calibrated from the primary standards are used for maintaining the radiometric scales and for calibrating working standards, the latter to be used by either NBS or other calibration laboratories throughout the country. The working standards are not always available from NBS, but they may be purchased from any of a number of industrial suppliers.

As seen in Chapter 1, the spectral radiance from a source is the energy per unit time, unit wavelength interval, and unit solid angle emitted in a specified direction by the total emitting surface of the source. The usual configuration for a lamp used as a standard of spectral radiance has a tungsten ribbon filament behind a plane quartz window, such as

Fig. 2.4 Photographs of various types of calibration sources (courtesy of the Eppley Laboratories): (A) standard of spectral radiance; (B) standard of spectral irradiance; (C) and (D) carbon filament and tungsten filament standards of total irradiance.

shown in Fig. 2.4A. A typical spectrum of radiation emitted by an incandescent tungsten filament at 2800°K is shown in Fig. 2.5. The strong continuum radiation from this lamp makes it a convenient standard for the visible and near-infrared regions, but at the operating temperature of tungsten very little of the energy is emitted in the ultraviolet.

The spectral irradiance on a surface is the energy per unit time and unit wavelength interval which is incident on a unit area of the surface. Incandescent tungsten is used also as a standard of spectral irradiance, but since the receiving surface is normally placed at a distance of 50 cm from the source, it is desirable for the standard of spectral irradiance to emit more energy than is necessary for a standard of spectral radiance. Consequently, the power rating for the former is normally about 1000 W, whereas that of the latter is only 100–150 W. In order to achieve the higher power in a small package, the configuration shown in Fig. 2.4B is often used. The lamp itself is of the quartz-iodine type with a coiled-coil filament.

Since the total irradiance on a surface is simply the amount of energy per unit time incident on a unit area of the surface, the spectral distribution of the energy is of little consequence. Thus, in principle, a line source could be used just as well as a continuum source as a standard of total irradiance. In practice, however, working standard lamps normally used for the purpose are of either the carbon-filament or tungsten-filament type, such as shown in Fig. 2.4C and 2.4D, respectively (Stair *et al.*, 1967). The power rating of carbon-filament lamps is normally in the 100–400 W range, while tungsten-filament working standards vary from 100 to as much as 5000 W in power output. Routine calibration of pyranometers is mainly based on the use of tungsten-filament lamps installed in large integrating spheres. The U.S. National Weather Service uses one 5000-W lamp (Hill, 1966), the Canadian Meteorological Service uses three 660-W lamps (Latimer, 1966), and the Eppley Laboratories use twelve 200-W and twelve 300-W lamps (Drummond and Greer, 1966), all of which are of the tungsten-filament type.

Arc Lamps The best known of the arc lights is the carbon arc, which is often used in spot lights, search lights, and other high intensity applications.

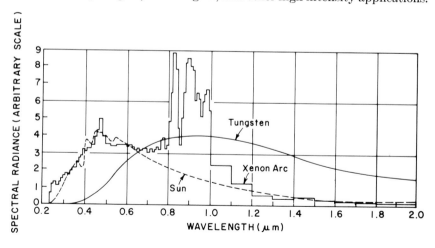

Fig. 2.5 Spectra of a high-pressure xenon arc lamp and an incandescent tungsten lamp superimposed on the extraterrestrial solar spectrum.

The spectrum is of the continuum type, and the extremely high temperature attainable in a carbon arc (3900–9000°K) results in high-energy emission in the ultraviolet region. The source of light in the carbon arc is a crater on the face of the positive electrode. Because of variations of the crater as the carbon burns, the source is difficult to control in both intensity and position, a fact which renders it unsatisfactory for purposes of precise calibrations.

High-pressure gaseous-discharge arc lamps suffer much less severe control problems than carbon arcs, although there is often some wandering of the arc. These high-pressure lamps, having as much as 50- to 70-atm internal pressure, yield extremely high radiance values from small areas. This is of importance in the creation of collimated beams. They have strong emissions in the ultraviolet. The most frequently used of the high-pressure lamps are the xenon arc, mercury arc, and xenon-mercury arc, although other types are available.

The output of the gaseous-discharge lamps is normally a strong line spectrum, with relatively little continuum radiation. However, the high pressure inside these arc lamps tends to broaden the lines, yielding a spectrum with some continuum and many broad peaks. An example of this pattern is the strong continuum in the visible and ultraviolet, with the band structure at longer wavelengths, shown for the xenon arc in Fig. 2.5. An equivalent color temperature of about 6000°K makes the xenon spectrum approximate that of the Sun at wavelengths below about 0.8 μm. For this reason, xenon arcs have found wide application in solar simulators developed in the space program. Mercury and other types of high-pressure arcs have less continuum radiation and a stronger band structure than that of xenon.

The possibility of explosion from these high-pressure arc lamps demands that safety precautions be exercised in their use. They must be housed in a rugged metal enclosure and provision must be made for stress-free expansion during warmup. Lamp terminals must be kept cool, normally by circulating water, during the period of lamp operation, and vertical mounting of the lamp is recommended.

Low-Pressure Discharge Lamps An electric discharge through a gas at low pressures excites the characteristic line emissions from the gas, and with negligible pressure broadening there is no significant continuum-type emission. Thus, low-pressure gaseous-discharge lamps are used mainly for wavelength calibrations. They are available with any of a number of gases, including argon, krypton, xenon, neon, and mercury, and in various physical configurations. The positions of selected lines emitted by the various gases are listed in Table 2.2.

TABLE 2.2

Positions in (μm) of Spectral Lines Emitted by Low-Pressure Gas-Discharge Lamps[a]

Argon	Krypton	Mercury	Neon	Xenon
0.39490	0.42740*	0.18491*	0.33699	0.46243
0.40444	0.43196*	0.19417	0.34477	0.46712*
0.41586*	0.43626	0.22622	0.35935	0.47342
0.41819	0.43761*	0.24820	0.53308	0.48070
0.42007*	0.44539	0.25635*	0.54006	0.82316*
0.42594	0.44637*	0.26520	0.58525	0.82801
0.43001	0.45024	0.28035	0.59448	
0.43336	0.55622	0.28936	0.60300	
0.69654	0.55703	0.30215	0.61431*	
0.70672	0.58709	0.31257*	0.62173	
0.72729	0.75874	0.31317	0.63048	
0.73840	0.76015*	0.33415	0.63344*	
0.75039	0.76852	0.36502*	0.63830*	
0.75146	0.76945	0.36544	0.64023*	
0.76351*	0.78548	0.36633	0.65065*	
0.77238	0.80595	0.40466*	0.65990	
0.79482	0.81044	0.40778	0.67170	
0.80062	0.81129	0.43475	0.69295	
0.81037	0.81901	0.43584*	0.70324*	
0.82645	0.82632	0.54607*	0.71739	
0.84082	0.82981	0.57696	0.74389	
0.84246		0.57907	0.75358	
			0.83776	

[a] The entries have been selected from a large number of lines, particularly for neon. The strongest lines are marked with an asterisk.

High-Temperature Blackbody Sources   Much precision radiometry, including the calibration of radiation receivers, determination of the Stefan–Boltzmann constant, and the evaluation of radiometric scales, has been accomplished by the use of high-temperature blackbody cavities. The usual configuration of such cavities is a blackened cone, cylinder, or sphere fitted with an aperture and heated by electrical means to temperatures of 1000 to 3000°K. The theory and use of such cavities has been discussed by Bedford (1970), Sparrow *et al.* (1962), Stair *et al.* (1960), and others. Emissivities are claimed to be as high as 0.999.

Although a number of blackbody furnaces are available commercially, their complexity, involving water cooling and large electrical currents, as well as their expense, has kept them from being popular as calibration sources for operational radiation instruments.

Two other sources which somewhat approximate blackbodies are the Globar and the Nernst glower. Both are normally in the form of rods, the former being about 5 cm long and 5 mm in diameter, and the latter about half that size. The operating temperatures are only about 1500 and 1800°K, respectively. These low temperatures combined with small sizes, the requirement for ballasted power supplies, and other shortcomings make them impractical for calibration purposes.

2.2.2 Low-Temperature Blackbody Sources

A number of blackbody sources suitable for the calibration of infrared radiometers have been developed, some high-precision sources to be used as radiation standards having been described by the National Bureau of Standards (Kostkowski *et al.*, 1970). Three such sources operate at the freezing points of metals (tin: 231.97°C; zinc: 419.58°C; gold: 1064.4°C) and have an emissivity of over 0.999. They have been adopted by the U.S. Department of Defense as standards of spectral radiance for the wavelength range 0.4 to 15 μm.

The space program has necessitated the use of large-area low-temperature blackbody sources (Hilleary *et al.*, 1968), two of which have been developed by Eppley Laboratories. The first of these, shown in the photograph of Fig. 2.6, has a source area of 65 cm^2 and operating temperature of -35 to $+50$°C (Karoli, 1970), and the second model has a source area of up to 750 cm^2 with a temperature range of -100 to $+70$°C. The blackbody itself,

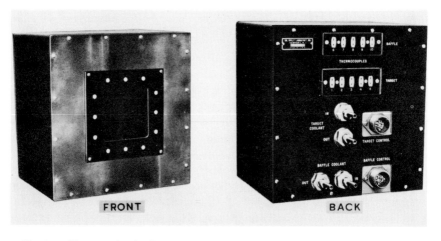

FRONT BACK

Fig. 2.6 Photograph of a large-area low-temperature blackbody calibration source (courtesy of the Eppley Laboratories).

which forms the back wall of a blackened cavity in each model, is of honeycomb construction with a depth of 2.5 cm and an emissivity said to be at least 0.995. The temperature of the wall is controlled to about 0.1°C. A similar low-temperature source was formerly available from EG & G, but its manufacture has been discontinued.

These low-temperature sources are suitable as radiance standards but not as irradiance standards. This means that they may be used to calibrate directional radiometers but not pyrradiometers, which receive energy from an entire hemisphere. This comes about because only the back wall of the cavity is temperature controlled. From information available, it appears that no low-temperature blackbody source is available commercially for calibrating pyrradiometers. It is, however, feasible to construct simple laboratory blackbody cavities for the purpose, as discussed in Chapter 11.

REFERENCES

Bedford, R. E. (1970). Blackbodies as absolute radiation standards. *In* "Advances in Geophysics" (A. J. Drummond, ed.), Vol. 14, pp. 165–202. Academic Press, New York.

Bolz, R. E., and Tuve, G. L., editors (1973). "Handbook of Tables for Applied Engineering Science," 2nd ed. C.R.C. Press, New York.

Brown, E. B. (1965). "Modern Optics." Von Nostrand-Reinhold, Princeton, New Jersey.

Drummond, A. J., and Greer, H. W. (1966). An integrating hemisphere (artificial sky) for calibration of meteorological pyranometers. *Sol. Energy* **10,** 7–11.

Hill, A. N. (1966). Calibration of solar radiation equipment at the U.S. Weather Bureau. *Sol. Energy* **10,** 1–4.

Hilleary, D. T., Anderson, S. P., Karoli, A. R., and Hickey, J. R. (1968). The calibration of a satellite infrared spectrometer. *Int. Astronaut. Congr. Proc. 18, Belgrade, 1967,* **2,** 423–437.

Hughes, A. L., and DuBridge, L. A. (1932). "Photoelectric Phenomena." McGraw–Hill, New York.

Jamieson, J. A., McFee, R. H., Plass, G. N., Grube, R. H., and Richards, R. G. (1963). "Infrared Physics and Engineering." McGraw–Hill, New York.

Karoli, A. R. (1970). Experimental blackbody (absolute) radiometry. *In* "Advances in Geophysics" (A. J. Drummond, ed.), Vol. 14, pp. 203–226. Academic Press, New York.

Kostkowski, H. J., Erminy, D. E., and Hattenburg, A. T. (1970). High-accuracy spectral radiance calibration of tungsten–strip lamps. *In* "Advances in Geophysics" (A. J. Drummond, ed.), Vol. 14, pp. 111–127. Academic Press, New York.

Kruse, P. W., McGlauchlin, L. D., and McQuistan, R. B. (1962). "Infrared Technology."
Wiley, New York.

Langley, S. P. (1900). The new spectrum. *Ann. Rep., Smithson. Inst.* pp. 683–692.

Latimer, J. R. (1966). Calibration program of the Canadian meteorological service.
Sol. Energy **10,** 4–7.

Smith, R. A., Jones, F. E., and Chasmar, R. P. (1968). "The Detection and Measurement
of Infrared Radiation," 2nd ed. Oxford Univ. Press (Clarendon), London and
New York.

Sparrow, E. M., Albers, L. U., and Eckert, E. R. G. (1962). Thermal radiation characteristics of cylindrical enclosures. *J. Heat Transfer* **84,** 73–79.

Stair, R., Johnston, R. G., and Halbach, E. W. (1960). Standard of spectral radiance for
the region of 0.25 to 2.6 microns. *J. Res. Nat. Bur. Std. Sect. A* **64,** 291–296.

Stair, R., Schneider, W. E., and Fussell, W. B. (1967). The new tungsten filament lamp
standards of total irradiance. *Appl. Opt.* **6,** 101–105.

Strong, J., and Lawrence, P. W., Jr. (1968). Bolometer theory. *Appl. Opt.* **7,** 49–52.

Thekaekara, M. P., Collingbourne, R. H., and Drummond, A. J. (1972). A comparison
of working standard pyranometers. *Bull. Amer. Meteorol. Soc.* **53,** 8–15.

Wolfe, W. L. (1965). "Handbook of Military Infrared Technology." Office of Naval
Research, Washington, D.C.

Zworkin, V. K., and Ramberg, E. G. (1947). "Photoelectricity." Chapman & Hall,
London.

CHAPTER THREE

Solar Radiation: Direct Component

Solar radiation reaches the top of the atmosphere at a mean rate, measured normal to the beam, of approximately 1.9 cal cm^{-2} min^{-1}. In spite of the fact that the Sun is about 1.39 million km in diameter, the visible disk (photosphere) subtends an angle at the Earth of only 0.545° of arc. On entering the atmosphere, the radiation is modified in intensity and polarization by scattering on molecules, water droplets, dust, and other aerosol particles in the atmosphere; by absorption by atmospheric gases and particulates; and by absorption and reflection by the underlying ground or water surface. Short-wave emissions, such as the airglow and aurora, from the atmosphere, although interesting phenomena in themselves, contribute a negligible amount to the overall energy budget.

The spectral distribution of solar radiation incident at the top of the atmosphere is compared with that emitted from a blackbody at a temperature of 6000°K in the two upper curves of Fig. 3.1. The significant discrepancies which occur in the ultraviolet region are mainly due to electronic transitions which occur in the overlying gases of the Sun. Beyond the regions of these transitions the two curves are similar in both shape and magnitude. The various atmospheric processes operate to change the spectral distribution as the radiation traverses the atmosphere to approximately that shown by the lower curve of Fig. 3.1. The main absorption is produced by water vapor, which is responsible for the several strong bands in the infrared region, and by high-altitude atmospheric ozone, which effectively limits radiation which reaches the ground in appreciable quantities

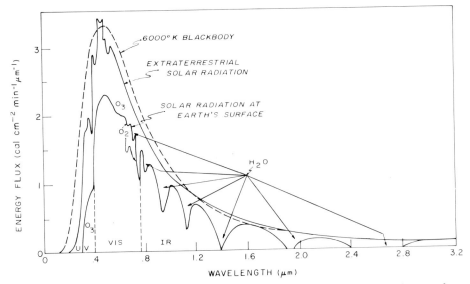

Fig. 3.1 Spectral distribution of solar radiation at the top of the atmosphere, and a typical distribution of that which reaches the Earth's surface, compared with the radiation emitted by a blackbody at a temperature of 6000°K.

to wavelengths greater than about 0.30 μm. Relatively small amounts of energy are absorbed by ozone and oxygen in the 0.6 to 0.7 μm region, and, although they are too weak to indicate in the diagram, there are some minor absorption bands of carbon dioxide in the near infrared. Scattering of radiation, which is particularly important at the shorter wavelengths, is mainly responsible for the decrease indicated by the curves in the visible and near-ultraviolet spectral regions.

Although scattering and absorption attenuate the solar energy during its downward traverse of the atmosphere, a relatively large part is transmitted directly and reaches the surface still as the approximately parallel beam one sees on looking at the disk of the Sun. The characteristics and measurement of this direct component are discussed in this chapter; the scattered component which reaches the surface from the sunlit sky is the subject of Chapter 4.

3.1 TRANSMISSION BY ATMOSPHERE

The incident solar beam is transmitted directly by the atmosphere according to the Bouguer–Lambert law (1760) which, for a plane-parallel and horizontally homogeneous atmosphere, can be written as

$$I_\lambda = I_{0\lambda} \exp(-\tau_\lambda \sec \theta_0) \tag{3.1}$$

Here $I_{0\lambda}$ and I_λ are the monochromatic intensities of the incident and transmitted radiations, respectively, τ_λ is the optical thickness of the atmosphere measured in the local zenith direction, and θ_0 is the angle between the local zenith and the direction of the Sun. By definition, the optical thickness of the whole atmosphere above an arbitrary height z is

$$\tau(\lambda, z) = \int_z^\infty B(\lambda, z) \, dz \tag{3.2}$$

where the attenuation coefficient $B(\lambda, z)$ is a function of both λ and z. Since the attenuation is produced by scattering and true absorption, we have $B = B_s + B_a$ and $\tau = \tau_s + \tau_a$, the subscripts indicating scattering and absorption, respectively.

3.1.1 Case of a Clear Atmosphere

It is convenient for purposes of discussion, to consider the effects of scattering and gaseous absorption separately, although it is realized that the two processes are not so readily separable in the actual atmosphere. The simplest atmospheric model is that of a nonabsorbing medium in which the scattering particles are all of a size much smaller than the wavelength, a criterion which applies principally to molecules of the atmospheric gases. For this case of a Rayleigh atmosphere, the volume scattering coefficient is given by

$$B^R(\lambda) = \frac{32\pi^3}{3\lambda^4} \frac{(n-1)^2}{N} \tag{3.3}$$

where n is the relative index of refraction of the medium and N is the number density of particles. This relation was first derived by Lord Rayleigh (1842–1919) in the early 1870's in connection with his famous explanation of the color and polarization of the light from the sunlit sky. An important point to notice in Eq. (3.3) is that the efficiency of scattering (attenuation) in the clear atmosphere is critically dependent on wavelength (λ^{-4}), in the sense that short wavelengths are scattered much more strongly than are long wavelengths. This fact, in combination with the spectral sensitivity of the eye and the spectral distribution of sunlight, accounts for the blue color we see in a clear daytime sky.

By knowing the distribution with altitude of the gases of the atmosphere, it is a simple matter to compute $B^R(\lambda)$ as a function of altitude, and perform the integration of Eq. (3.2) to determine the Rayleigh optical thickness τ^R of the air above any selected level in the atmosphere. The results of these computations are shown in Fig. 3.2 by curves of τ^R vs λ, for 16 altitudes up to 32 km for the (1959 A.R.D.C.) standard atmosphere. By entering values of τ^R taken from these curves into Eq. (3.1), we can compute the attenuation of solar radiation which is produced by all of the Rayleigh atmosphere above a selected level. Similarly, by knowing the extra-atmosphere value $I_{0\lambda}$ of solar radiation intensity, we can determine

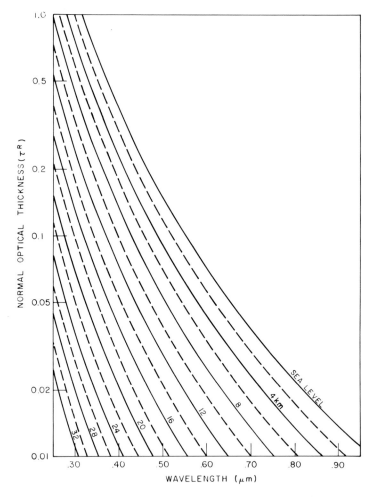

Fig. 3.2 The normal optical thickness of a Rayleigh atmosphere as a function of wavelength for various altitudes above the surface.

TABLE 3.1

Rayleigh Optical Thickness τ^R at Sea Level ($z = 0$) for Various Wavelengths

λ (μm)	τ^R	λ (μm)	τ^R	λ (μm)	τ^R
0.25	2.74	0.55	0.101	0.85	0.0173
0.30	1.25	0.60	0.0708	0.90	0.0138
0.35	0.650	0.65	0.0512	0.95	0.0111
0.40	0.373	0.70	0.0379	1.00	0.00900
0.45	0.229	0.75	0.0287		
0.50	0.149	0.80	0.0221		

the intensity I_λ of the direct solar beam at the selected altitude in the Rayleigh atmosphere. Values of τ^R for the entire atmosphere ($z = 0$) are listed for several wavelengths in Table 3.1.

The very strong wavelength dependence is demonstrated, the value of τ^R being more than 300 times as great at $\lambda = 0.25$ μm as at $\lambda = 1.00$ μm.

3.1.2 Case of a Turbid Atmosphere

There is ample evidence that there are sufficient numbers of dust, haze, and other types of non-Rayleigh particles in even the clearest cases of the natural atmosphere to produce important radiative effects. For instance, measurements by Eddy (1961) of the solar aureole in a clear atmosphere, show significant aerosol effects up to altitudes of at least 80,000 ft (24.4 km). From measurements of the scattering and absorption coefficients as a function of altitude in the atmosphere over England, Waldram (1945) found an extreme variability of aerosol effects in both space and time. Vertical profiles of the volume scattering coefficient for selected cases of clean air are shown by the curves of Fig. 3.3. The measurements were in "white" light, and therefore represent an integration over the visible spectrum. The pure air curve is for a Rayleigh atmosphere at 0.54 μm. A large variability occurs, even for a clean atmosphere, and the actual volume scattering is several times larger than that for the Rayleigh atmosphere. The curve for 5–9–42 is for a case of considerable haze extending to high altitudes.

Typical values of the volume attenuation coefficient at low levels in the atmosphere, as measured by Waldram (1945) and as computed for realistic aerosol models by Fraser (1959), are compared in Fig. 3.4 with those for a Rayleigh atmosphere at sea level. Models A and D represent a maritime aerosol, and model C is for a continental-type aerosol. The data show that for both the measurements and the atmospheric models con-

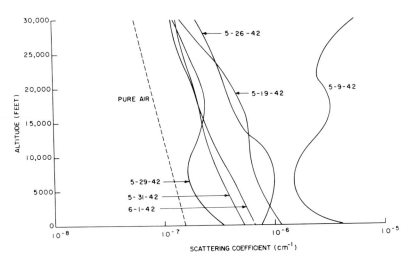

Fig. 3.3 The volume scattering coefficient vs altitude as observed in "white" light on different occasions by Waldram (1945).

sidered, aerosol scattering and absorption are more effective in attenuating radiation than is scattering by Rayleigh particles, at least at wavelengths $\lambda > 0.5$ μm. The relative difference would be less at shorter wavelengths because the scattering by aerosols is much less dependent on wavelength than is that of Rayleigh scattering.

In recognition of the radiative effects of aerosols, absorbing gases, and other atmospheric constituents several methods have been developed for using atmospheric attentuation of solar radiation as an index of atmospheric turbidity. The two main methods of characterizing turbidity conditions of the atmosphere, both of which were developed during the 1920's, are discussed below.

Linke Turbidity Factor   The first of the standard methods of turbidity characterization is that developed by Linke (1922, 1929). Linke considered that a logical unit of attenuation is that of a Rayleigh atmosphere, and defined a turbidity factor T as the number of Rayleigh atmospheres required to produce a measured amount of attenuation. The original method was based on measurements of the total (wavelength-integrated) radiation in the direct solar beam, but in a later (and somewhat improved) method the problem of variable amounts of water-vapor absorption is partially alleviated by confining the measurements to energy at wavelengths less than 0.63 μm. The principles are the same, however. The flux of energy in the direct solar beam, which reaches a measuring instrument

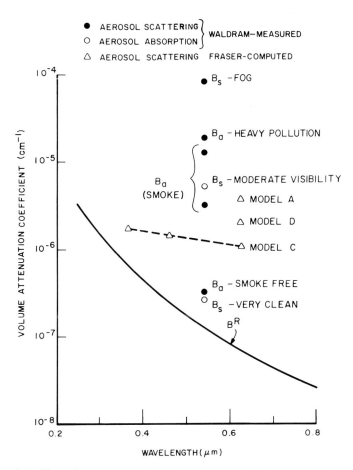

Fig. 3.4   The volume attenuation coefficient as a function of wavelength for a Rayleigh atmosphere at sea level, for various aerosol models as computed by Fraser (1959), and as measured in the low levels of the atmosphere by Waldram (1945).

at a given time, is given by the Bouguer–Lambert law as

$$F = (1/S) \int_0^\infty F_0(\lambda) e^{-\tau(\lambda, m)m} \, d\lambda \qquad (3.4)$$

Here $F_0(\lambda)$ is the spectral flux of solar radiation at the top of the atmosphere, m is relative air mass,* and the parameter S, introduced for normal-

* For a plane parallel atmosphere, $m = \sec \Theta_0$ (solar zenith angle), which is in error by less than 2% for the actual atmosphere for Θ_0 less than 78°.

izing the Sun–Earth distance to its mean value, is given in terms of the actual and mean distances, R and R_m, respectively, by $S = R^2/R_m^2$.

The exponential factor $\tau(m)$ is composed of optical thicknesses due to Rayleigh scattering $\tau^R(\lambda)$, aerosol extinction $\tau^A(\lambda)$, and water-vapor absorption $\tau^W(\lambda)$. Then the Linke turbidity factor is defined by the relation

$$F = (1/S)F_0 e^{-[T\bar{\tau}^R(m)m]} \tag{3.5}$$

where $\bar{\tau}^R(m)$ is the mean value of τ^R weighted according to the distribution of transmitted energy and integrated over all wavelengths. From Eq. (3.5) we can write

$$T = P(m)(\log F_0 - \log F - \log S) \tag{3.6}$$

where

$$P(m) = [m\bar{\tau}^R(m) \log e]^{-1} \tag{3.7}$$

By determining $\bar{\tau}^R(m)$ by wavelength integrations for various values of m, it is possible to determine $P(m)$ as a function of m, and apply it to all cases. The best numerical values of $P(m)$ available (CSAGI, 1958) are given in Table 3.2. From those values, in combination with measurements of F and computations of F_0 and S, it is a simple matter to determine T by the use of Eq. (3.6).

For stations at which the pressure p is considerably different from the standard $p_s = 1000$ mb, the above analysis can be corrected to yield an "extrapolated turbidity factor T_p" for the station by the relation (Feussner and Dubois, 1930)

$$T_p = 1 + (T - 1)[P(m_z)/P(m)] \tag{3.8}$$

where $P(m_z)$ is the value of $P(m)$ at (pressure) altitude z. The usual recommendation is that T_p be used in place of T for stations at which the pressure differs from 1000 mb by more than 50 mb (CSAGI, 1958).

TABLE 3.2

Factor $P(m)$ for the Computation of the Linke Turbidity Factor T from Measurements of Total Radiation

m	$P(m)$	m	$P(m)$	m	$P(m)$	m	$P(m)$
0.5	43.9	2.5	10.76	5.5	6.06	8.5	4.60
0.6	37.0	3.0	9.35	6.0	5.72	9.0	4.45
0.8	28.4	3.5	8.33	6.5	5.43	9.5	4.31
1.0	23.2	4.0	7.55	7.0	5.18	10.0	4.19
1.5	16.3	4.5	6.95	7.5	4.96		
2.0	12.9	5.0	6.45	8.0	4.77		

The Linke turbidity factor has been found useful for comparisons of atmospheric turbidity under different conditions, but it suffers from one serious difficulty. Measurements have shown that T varies with m even when atmospheric conditions are unchanged, the result being a fictitious diurnal variation of turbidity. The reason for this is that the wavelength dependencies of water-vapor absorption and aerosol absorption and scattering are quite different from that of Rayleigh scattering. Efforts to derive a modified form of T based on a constant value of atmospheric water-vapor have not been particularly successful.

As mentioned above, one method of minimizing water-vapor effects in determining turbidity is to compute T from measurements of radiation at wavelengths below the main water-vapor bands. The most general practice has been to make the measurements by use of a pyrheliometer with and without the red cutoff filter Schott RG 2. By this means, the energy of the solar beam for all wavelengths $\lambda < 0.630$ μm is measured, and by a redetermination of F_0 and $P(m)$ it is possible to use Eq. (3.6) for computing T for this short-wavelength region of the spectrum. This procedure, however, doubles the number of measurements necessary, demands higher total accuracy, and requires careful determinations of transmission characteristics of the cutoff filter together with air-mass dependent corrections for nonstandard filter transmission characteristics. Furthermore, a temperature measurement of the filter itself is required for large temperature changes, as the wavelength cutoff of glass absorption filters varies with temperature. Thus the very attractive feature of simplicity is lost in this short-wavelength method, and it has received only limited acceptance in practice. The determination of turbidity by means of the Ångström turbidity coefficient is usually preferred if pyrheliometric measurements with filters are available.

Ångström Turbidity Coefficient In order to take account of the difference in transmission characteristics between aerosols and Rayleigh particles, A. K. Ångström (1929, 1930, 1961) represented the normal optical thickness $\tau^A(\lambda)$ due to aerosols in terms of a "turbidity coefficient" β and a wavelength exponent α, as

$$\tau^A(\lambda) = \beta\lambda^{-\alpha} \tag{3.9}$$

Then Eq. (3.4) can be expressed as

$$F = (1/S) \int F_0(\lambda) \exp \{- [\tau^R(\lambda, m) + \tau^A(\lambda, m)]m\} \, d\lambda \tag{3.10}$$

if absorption by water vapor and ozone is neglected for the moment. By this relation, together with measurements of F and computations of S,

$F_0(\lambda)$, and $\tau^R(\lambda, m)$, we can determine $\tau^A(\lambda, m)$. Further, if α has a known value, then a determination of β is immediate.

Considerable effort has been spent by A. K. Ångström (1961), Schüepp (1949), and others in determining the magnitude and variability of α. From a theoretical standpoint, the limits of α should be approximately 4 for very small particles and 0 for very large particles. The extensive work on observations in the natural atmosphere indicate that a good average value is $\alpha = 1.3 \pm 0.2$, with practical limits about 0.5 and 1.6 (A. K. Ångström, 1961).

The problem of the optical thickness $\tau^W(\lambda, m)$ due to absorption of radiation by water vapor can be largely avoided for determinations of β and α by measuring the energy in only the visible and ultraviolet spectral regions (Linke's *Kurzstrahlung*), since most water-vapor absorption bands are in the near infrared. These short-wave measurements* are usually made by a pyrheliometer together with one of the Schott absorption filters OG1, RG2, or RG8, the lower cutoff wavelengths of which are standardized (by WMO) at 0.525, 0.630 and 0.700 μm, respectively. By measurements of the energy at $\lambda < 0.525$ μm and that at $\lambda < 0.630$ μm it is theoretically possible to get concurrent determinations of both β and α.

Ozone absorption, which occurs in the Hartley–Huggins bands and Chappuis bands, is taken into account by increasing $\bar{\tau}^R$ appropriately for mean ozone amounts. Errors due to differences between actual and mean ozone amounts would be only second order and of minor significance.

From a given extraterrestrial solar energy spectral distribution and under the assumption of a constant value of α, it is possible to compute, from Eq. (3.10), the amount of energy in a given band which would be transmitted through the atmosphere for various values of β and m. The results obtained from these calculations for the wavelength region $\lambda < 0.630$ μm, defined by the RG2 cutoff filter, and with the Nicolet distribution (solar constant 1.98 cal cm^{-2} min^{-1}), are shown by the curves of Fig. 3.5 for values of air mass m from 1 to 6. The values of solar irradiance are with respect to the International Pyrheliometric Scale 1956. By entering the diagram with the measured flux F_k for $\lambda < 0.630$ μm and the absolute airmass m, the turbidity coefficient β can be read off from the family of curves. Similar curves for the RG8 filter are given by A. K. Ångström (1970).

The data of Fig. 3.5 are applicable for a location at which the pressure

---

* Such short-wave determinations are actually made by subtracting from the total flux F_t measured by a pyrheliometer without a filter the longwave flux F_R measured by the pyrheliometer with the cutoff filter. Then the short-wave flux is $F_k = F_t - D_1 F_R$ where D_1 is the reciprocal of the wavelength-integrated transmittance of the filter.

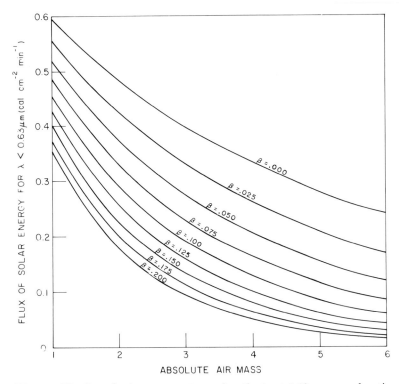

Fig. 3.5   The flux of solar energy at wavelengths $\lambda < 0.63$ μm as a function of air mass for various values of the Ångström turbidity coefficient.

is 1000 mb. If the actual pressure is outside the range of approximately 970 mb $< p <$ 1030 mb, then the value of β obtained from the diagram must be multiplied by the factor $p/1000$, to yield a corrected turbidity coefficient β_p for the given location.

Since there are small variations in the lower cutoff wavelengths λ_c (wavelengths at which the transmission is 50%) of different specimens of the OG1, RG2, RG8, and similar absorption filters, it may be necessary to add a correction ΔF_k to the measured value of flux F_{km} to reduce it to what would have been measured with a filter of standard cutoff wavelength λ_s (0.525 μm for OG1, 0.630 μm for RG2, etc.). If $\Delta\lambda$ is given by $\Delta\lambda = \lambda_c - \lambda_s$, then the corrected short-wave flux F_k is

$$F_k = F_{km} \pm \Delta F \tag{3.11}$$

where the plus $(+)$ sign applies for $\lambda_c > \lambda_s$, and vice versa. The additive correction is highly variable with air mass m and turbidity coefficient β, and for accurate results it cannot simply be included in the filter factor as has been a relatively widespread practice. In Table 3.3 are given values of the additive term ΔF for filters OG1 and RG2 corresponding to $\Delta\lambda = 0.01$ μm and various values of β and m (Ångström and Drummond, 1961). The tabulation can be made applicable to values of $\Delta\lambda \neq 0.01$ μm by multiplying the entries by the factor $\Delta\lambda/0.01$, where $\Delta\lambda$ is in micrometers. The procedure for determining the turbidity coefficient β with the nonstandard filter is thus: (a) determine from Fig. 3.5 (or the equivalent) a preliminary value of β; (b) from Table 3.3 determine ΔF for the air mass and the preliminary β; and (c) obtain a corrected F_k by adding (or subtracting) ΔF to F_{km}, by which a final value of β is determined from Fig. 3.5.

In 1970 the World Meteorological Organization recommended that turbidity measurements be made at designated Air Pollution Stations. Since that time the Environmental Protection Agency of the U.S. Government has established over 80 such stations, of which 50 are in the United States (Bilton *et al.*, 1973). The measurements are made by means of a photometer (normally designated a Sun photometer) based on the design originally developed by Volz (1959). The latest model, and the one used in most of the turbidity network stations, is a dual-channel device for making measurements of the direct solar radiation in narrow spectral regions centered at 0.38 and 0.55 μm. The acceptance aperture is a 1°–2° half-angle cone, and the detector is a photovoltaic selenium cell. The signal from the detector is amplified by a small operational amplifier and read directly by an attached microammeter.

The instrument is a self-contained unit of dimensions approximately $8 \times 9 \times 20$ cm and weighing less than 1 kg. Measurements are usually made by holding the instrument by hand and orienting it toward the Sun

TABLE 3.3

Values of ΔF (in cal cm^{-2} min^{-1}) for Filters OG1 and RG2 Corresponding to a Lower Cutoff Wavelength Shift $\Delta\lambda = 0.01$ μm for Specified Values of Turbidity Coefficient β and Air Mass m^a

Filter OG1					Filter RG2				
β/m	1.0	2.0	3.0	4.0	β/m	1.0	2.0	3.0	4.0
0.00	0.026	0.023	0.020	0.018	0.00	0.024	0.022	0.021	0.020
0.05	0.023	0.018	0.014	0.011	0.05	0.022	0.019	0.016	0.014
0.10	0.020	0.014	0.010	0.007	0.10	0.020	0.015	0.012	0.009
0.20	0.016	0.009	0.005	0.003	0.20	0.017	0.011	0.006	0.004

[a] Ångström and Drummond, 1961.

with the aid of a pinhole target diopter attached to the side of the case. The diopter is fitted with a scale of angles and can be used in conjunction with a spirit level for obtaining a direct measure of relative air mass m at the time of observation, although it is often more convenient to determine m from computed elevations of the Sun. Calibration of the network instruments is carried out by comparison with standard instruments maintained for the purpose by the Environmental Protection Agency.

In order to determine the Ångström turbidity coefficient β and wavelength exponent α [see Eq. (3.9)], we expand Eq. (3.10) by including the optical thickness of ozone, which cannot be neglected at $\lambda = 0.50$ μm, and obtain the relation

$$F(\lambda) = (1/S)F_0(\lambda) \exp\{-[\tau^R(\lambda) + \tau^O(\lambda) + \tau^A(\lambda)]m\} \quad (3.12)$$

We have omitted a wavelength integration for this quasimonochromatic radiation. Values of τ^R applicable at sea level are approximately 0.195 and 0.063 for wavelengths 0.38 and 0.50 μm, respectively, and those for τ^O under normal ozone amounts are 0.0 and 0.005. On introducing these, together with the measurements of F and known values of F_0, S, and m, into Eq. (3.12), $\tau^A(\lambda)$ can be determined for the two wavelengths. Then by applying Eq. (3.9) to the two cases and taking the ratio, we have

$$\alpha = \left\{\ln\left[\frac{\tau^A(\lambda = 0.38)}{\tau^A(\lambda = 0.50)}\right]\right\} \Big/ \left\{\ln\left[\frac{0.50}{0.38}\right]\right\} \quad (3.13)$$

Once α is determined, the turbidity coefficient for each wavelength can be computed directly from Eq. (3.9).

Data from the existing turbidity network are collected and published by the Environmental Data Service, Asheville, North Carolina.

3.1.3 Case of a Cloudy Atmosphere

Clouds are very strong attenuators of direct solar radiation as is obvious from the fact that the disk of the Sun is obscured by all but thin cirrus and very thin altostratus clouds. Although a significant fraction of the incident sunlight is transmitted by a dense cumulus or a heavy overcast of stratus, the transmission is mainly by multiple scattering on the water droplets, and is not a direct transmission of the original beam.

To a good approximation, the water droplets which constitute a cloud or fog can be considered as transparent spheres, in which case all of the radiation extracted from the original beam reappears as scattered energy. Furthermore, from a knowledge of the size-frequency distribution of the droplets and the liquid-water content of the cloud, the attenuation coefficient of the cloud can be computed by use of the Mie theory of scattering. For a given droplet radius r and a number concentration N of the droplets, the attenuation coefficient B^c for the cloud is given by

$$B^c = Nr^2f(\lambda/r) \tag{3.14}$$

where $f(\lambda/r)$ has an almost constant value of approximately 2 for visible light and droplet radii typical of clouds. For natural clouds, r is by no means constant, in which case an integration of Eq. (3.14) over the size-frequency distribution of the cloud droplets is required. Deirmendjian (1969) applied the complete Mie theory to a model of the droplet distribution for a cumulus cloud and obtained the values listed in the following tabulation for the attenuation coefficient at the indicated wavelengths.

$\lambda(\mu m)$	B^c (cm^{-1})
0.45	1.633×10^{-4}
0.70	1.673×10^{-4}
1.19	1.729×10^{-4}
1.45	1.763×10^{-4}
1.61	1.758×10^{-4}
1.94	1.805×10^{-4}
2.25	1.836×10^{-4}
3.90	2.064×10^{-4}

Measurements of the attenuation coefficient in clouds have been reported by Waldram (1945), Aufm Kampe (1950), and others. Mean values for different cloud types obtained by Aufm Kampe (1950), with an airborne transmissometer using visible light, are given in the following tablulation.

Cloud type	B^c (cm^{-1})
Cumulus congestus	1.95×10^{-3}
Fair weather cumulus	9.77×10^{-4}
Stratocumulus	3.91×10^{-4}
Stratus	2.79×10^{-4}
Altostratus	2.67×10^{-4}

These values for cumulus-type clouds are up to an order of magnitude higher than those computed by Deirmendjian. If we neglect errors of measurement, which may be large over the short pathlengths used by Aufm Kampe, the discrepancies may be explained on the basis of the

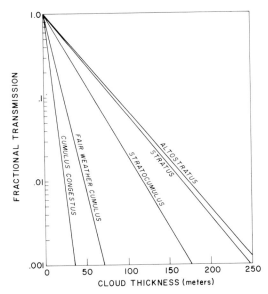

Fig. 3.6 The fractional transmission of direct solar radiation as a function of cloud thicknesses for various types of clouds.

difference between the theoretical model of Deirmendjian and the actual clouds in which the measurements were made. Aufm Kampe (1950) gives the average drop radius as 9.5 μm and the liquid-water content as 2.5 g m^{-3} in a cumulus congestus cloud, whereas the corresponding values used by Deirmendjian are 4.0 μm and 0.063 g m^{-3}.

The fractional transmission of the direct solar beam as a function of cloud thickness for the different types of clouds is shown in Fig. 3.6. The curves are computed from the Bouguer–Lambert law by the use of the attenuation coefficients measured by Aufm Kampe (1950). The diagram demonstrates that a 99% attenuation of the direct beam is produced in 170 m of a stratus or altostratus cloud, and in only about 25 m of a cumulus congestus cloud. This explains very well the obscuration of the disk of the Sun by relatively thin natural clouds.

3.2   MEASUREMENT OF THE DIRECT COMPONENT: PYRHELIOMETERS

3.2.1   Historical Background

The measurement of direct solar radiation received considerable emphasis from the time Pouillet devised what was apparently the first pyrheliometer (about 1837) until well into the twentieth century, and pyrheliometry reached a high state of development during that hundred-year period. The main motivation for solar radiation measurements during the first quarter of the twentieth century was for the establishment of the value of the solar constant, and the time variations of that value. Once that work was put on a routine basis, there was a waning interest in the development of pyrheliometers, even though considerable improvement was desirable. Only during the last decade or so has the interest been revived, the revival owing no doubt to a combination of the availability of new techniques in instrumentation and the need for improved radiation data for energy budget studies and space exploration. The advanced instrumentation has been confined mainly to experimental programs so far and has not been incorporated into radiation measurement networks for routine observations. In fact, the standard pyrheliometers against which operational instruments have been calibrated, viz., the Ångström electrical compensation pyrheliometer, the Abbot silver-disk pyrheliometer, and the Smithsonian water-flow pyrheliometer, were developed at about the turn of the century. A large number of devices for measuring the direct radiation from the Sun have been developed over the last 150 years or so, of which only a few are still in use. Some of the most interesting of the historical instruments are briefly described below.

Herschel's Pyrheliometer (Actinometer)   Herschel's actinometer, developed about 1825, was one of the first instruments used for measuring the heating effect of direct solar radiation, and it was the first in which the rate of cooling was introduced into solar radiation measurements (Whipple, 1915). The Herschel instrument consisted of a thermometer with a large cylindrical glass bulb filled with a liquid which was colored deep blue to promote absorption of solar radiation. In operation, the thermometer bulb was shaded from direct sunlight for 1 min, then exposed for 1 min, and again shaded 1 min. The thermometer readings obtained during this three-stage cycle could, on calibration by a known energy source, be interpreted in terms of the direct solar energy on the instrument. The method of making radiation determinations by alternately shading and exposing the sensor was adopted in many of the later instruments, and it is still used, for example, in the Abbot silver-disk and Michelson pyrheliometers.

Hodgkinson's Pyrheliometer (Actinometer)   The principle of Hodgkinson's actinometer is the same as that of Herschel's actinometer, but it was made lighter and more sturdy for use in the mountains. The 2.5-cm-diam thermometer bulb was filled with alcohol colored with aniline blue (Anon., Sixth Annual Exhibition of Instruments, 1885). The thermometer bulb was mounted in the middle of a bright metal tube of about 6 cm in diameter and 45 cm in length, the ends of which were covered by caps of clear plate glass. The stem of the thermometer projected through a hole in the wall of the tube. As with Herschel's actinometer, the reading of the thermometer, after a series of alternate intervals of shading and exposure of the bulb, was used (after calibration) to determine the flux of solar radiation.

Crova Pyrheliometer (Actinometer)   Crova's actinometer was also similar to that of Herschel, and it was used in a similar manner (Whipple, 1915). The 2-cm-diam thermometer bulb was filled with alcohol, but the index was of mercury, for ease of reading. Absorption of solar radiation was promoted by blackening the bulb itself, instead of coloring the alcohol as both Herschel and Hodgkinson had done. Radiation was introduced through a 10-mm aperture in the surrounding metal envelope. A series of diaphragms was arranged to assure that entering radiation fell on the bulb and to decrease the effect of wind on the instrument.

Pouilett's Pyrheliometer   The first successful research on the solar constant appears to have been that of Pouillet in 1837 (Glazebrook, 1923). For that work he used an instrument for which he introduced the term "pyrheliometer." The radiation receiver was a flat cylindrical vessel which was filled with water (mercury at a later date), and acted as a calorimeter. The exposed face of the cylinder was blackened to provide an absorbent surface, while the other surfaces of the cylinder were silvered to decrease radiative exchanges. A thermometer bulb was sealed into the liquid of the cylinder, and the stem of the thermometer extended along the axis. Rotation of the cylinder around this axis mixed the liquid to provide more representative temperature measurements.

In operation, the blackened face of the receiver was directed to a part of the sky away from the Sun for a period of 5 min, and the fall of temperature was noted. During the next 5 min the black surface was exposed normally to the solar beam, and the rise of temperature noted. Finally, the receiving surface was again directed away from the Sun for 5 min and the fall of temperature was noted. From a knowledge of the heat capacity of the calorimeter, the total heat received per unit surface and unit time could be computed. Pouillet obtained a value (1.7 cal cm^{-2} min^{-1}) of the solar constant which was within 14% of our present most generally accepted value (1.94 cal cm^{-2} min^{-1}).

Violle's Pyrheliometer (Actinometer)   Violle's pyrheliometer (Whipple, 1915) consisted of a thermometer bulb mounted in the center of a double-walled sphere, the space between the walls being filled with a constant temperature ice-water bath. The thermometer bulb was shaded or exposed to the solar beam by the operation of a shutter which closed or opened the instrument aperture. In operation, the shutter was closed until an equilibrium temperature was obtained. Then the shutter was opened and the rate of rise of temperature measured. By knowing the heat capacity of the thermometer bulb, the rate of energy absorption could be computed.

The Violle actinometer was employed for a considerable time in Europe. Vallot used the instrument for measuring the solar constant from the top of Mont Blanc in 1887, thereby obtaining a value close to the currently accepted figure.

Marvin Pyrheliometer   It is interesting that the pyrheliometer designed by C. F. Marvin about 1910 (Covert, 1925) is only one of his contributions to radiation instrumentation. He had already designed a photographic sunshine recorder, which was used by the U.S. Weather Bureau from 1891 to 1905, and apparently cooperated with D. T. Maring in the development of a thermoelectric sunshine recorder about 1893. Marvin was Chief of the U.S. Weather Bureau from 1913 to 1934.

The design of the Marvin pyrheliometer is in many respects similar to that of the Smithsonian Abbot silver-disk pyrheliometer. The sensing element in both cases is a blackened silver disk, and the method of opera-

Fig. 3.7 The Marvin pyrheliometer (far right) with various accessories used in measuring the flux of direct solar radiation (after Covert, 1925).

tion is essentially the same. The main difference between the two is that the mercury thermometer of the Smithsonian instrument is replaced in the Marvin pyrheliometer, by a resistance wire for temperature sensing.

The configuration of the Marvin pyrheliometer can be seen from the photograph of Fig. 3.7. The sensing element was a disk of blackened silver 4.5 cm in diameter and 0.3 cm in thickness, which is suspended by three fine wires in a metal support encased in the spherical wooden shell which formed the base of the instrument. The wood of the shell protected the disk from short period changes of the ambient environment. Radiation was admitted to the instrument through a collimator tube fitted with optical diaphragms and blackened inside, and the instrument was pointed at the Sun by a clock-driven equatorial mount. An electrically operated shutter mounted in front of the entrance aperture caused alternate 1-min intervals of shading and exposure of the blackened surface of the disk.

Because of its simpler operation, the Marvin pyrheliometer replaced, in 1914, the last of the Ångström pyrheliometers used for solar normal incidence measurements by the Weather Bureau. Four Marvin pyrheliometers were in use in the United States radiation network in 1925 (Covert, 1925) and two were still in use a decade later (Hand, 1937). As far as is known, the Marvin instrument is now completely obsolete.

Dorno's Pyrheliometer (Pyrheliograph)   One of the many devices for measuring radiation which have been developed by the Davos Physical-Meteorological Observatory was the pyrheliometer of C. Dorno (1922). A photograph of the instrument is shown in Fig. 3.8. A blackened thermopile consisting of 18 copper-constantan thermoelements was placed in the center of a cylindrical cavity 8 mm in diameter and 24 mm in length recessed into a massive 3-kg cylinder of copper. As in the Marvin and Michelson pyrheliometers, the copper provided a large heat capacity which maintained an even temperature of the inside cavity and decreased perturbations from the environment. The walls of the cavity in front of the thermopile were polished to reflect stray radiation to the blackened thermopile surface, while the rear portion of the cavity was itself blackened to absorb radiation which bypassed the thermopile. Radiation entered the instrument through a quartz lens of 20 mm in diameter and 86 mm in focal length. The receiving surface of the thermopile was 2.8 mm behind the focal point of the lens. By use of exchangeable diaphragms at the focal plane three different acceptance apertures could be obtained. The smallest of these was only about three-quarters of a degree in angular diameter, which means that only about 8 arc min of sky around the edge of the solar disk was seen by the instrument. The other two diaphragms gave angular apertures of about 1.1° and 2.1°. Even the largest of these apertures is

Fig. 3.8 The pyrheliometer of C. Dorno developed at the Physical-Meteorological Observatory, Davos, Switzerland (after Dorno, 1922).

much smaller than apertures of the most popular pyrheliometers in present use (Abbot silver disk and Eppley: 5.7° cone; Kipp and Zonen: 10.2° cone; Ångström: 3 × 6° to 6 × 24° rectangle). Thus, measurements from the Dorno instrument would be less subject to errors arising from scattered light in the circumsolar zone of the sky, but high-precision tracking of the Sun was required.

Although the Dorno pyrheliometer appears to have incorporated some superior features, it was never widely accepted and it has been completely superseded by other types of instruments.

Abbot's Balloon Pyrheliometer   During the solar constant work of the Smithsonian Institution, it was suggested by A. K. Ångström that an instrument be devised by which the solar energy could be measured at the highest altitude reached by a balloon. A design for such an instrument was developed, and five copies of the automatically recording balloon pyrheliometer were built and flown from Catalina Island, California during 1913. All were recovered, modified appropriately, and recalibrated; three were flown again from Omaha, Nebraska, in July 1914 (Abbot *et al.*, 1922). The highest altitude at which solar radiation measurements were obtained was 24 km on 11 July, at which altitude the measured solar radiation was 1.84 cal cm^{-2} min^{-1}. A 2% correction for scattering and absorption by the atmosphere above 24 km was considered reasonable, so, after correction to the mean Sun–Earth distance, the balloon flights corroborated the solar constant value of 1.93 cal cm^{-2} min^{-1}, already advanced by the Smithsonian Institution.

3.2.2 Classification of Pyrheliometers

Following the Commission for Instruments and Methods of Observation of the World Meteorological Organization (1965), we classify pyrheliometers as standard, first class, and second class, in accordance with the criteria given in Table 3.4.

On the basis of these criteria the commercially available pyrheliometers are classified as follows:

Standard pyrheliometers
  Ångström electrical compensation pyrheliometer
  Abbot silver-disk pyrheliometer
First-class pyrheliometers
  Michelson bimetallic pyrheliometer
  Linke–Feussner iron-clad pyrheliometer
  New Eppley pyrheliometer (temperature compensated)
  Yanishevsky thermoelectric pyrheliometer

TABLE 3.4

Classification of Pyrheliometers

	Standard	1st class	2nd class
Sensitivity (mW cm^{-2})	± 0.2	± 0.4	± 0.5
Stability (% change per year)	± 0.2	± 1	± 2
Temperature (maximum error due to changes of ambient temperature—%)	± 0.2	± 1	± 2
Selectivity (maximum error due to departure from assumed spectral response—%)	± 1	± 1	± 2
Linearity (maximum error due to nonlinearity not accounted for—%)	± 0.5	± 1	± 2
Time constant (maximum)	25 sec	25 sec	1 min

Second-class pyrheliometers

   Moll–Gorczynski pyrheliometer
   Old Eppley pyrheliometer (not temperature compensated)

   The Smithsonian water-flow pyrheliometer was omitted from the list of standard instruments, but it has been one of the primary standards, especially in the United States, against which silver-disk pyrheliometers, and, in some cases, Ångström pyrheliometers, were calibrated. The various types of instruments are discussed individually in the sections which follow.

3.2.3 Primary and Standard Pyrheliometers

Smithsonian Water-Flow Pyrheliometer This instrument was developed by the Smithsonian Institution as a primary standard for radiation measurements; the construction was started in 1903 and completed about 1910. As first designed, the instrument was essentially a single-cell calorimeter in which the walls of a chamber at the lower end of a collimating tube were coated with a highly absorbing material, such as a lampblack, in which the direct solar radiation, plus that from the sky in a small solid angle surrounding the solar disk, was absorbed. The outside of the walls of the cell were bathed by a circulating liquid, first nitrobenzol and later distilled water, by which the energy was carried away from the chamber by the flow of liquid at the same rate it was added by the absorption of radiation, once steady-state conditions were established. The rate of removal was determined by measuring the temperature change of the liquid at the inlet and outlet ports, and also the rate of flow.

   The original model of this single-chamber water-flow pyrheliometer gave stable and reproducible results. It was used as the primary standard for the calibration of secondary instruments, principally the silver-disk pyrheliometer, until an improved double-chamber model was developed and put into operation in 1932 (Abbot and Aldrich, 1932). The addition of the second chamber, first suggested by W. M. Shulgin of the Timiriaseff's Academy of Rural Economy of Moscow (1927), significantly improved the performance of the instrument, and it was employed by Abbot at Mount Wilson and Table Mountain in deriving the 1932 standard scale of radiation. Abbot and Aldrich (1932) claimed an accuracy of ±0.2% in individual energy determinations.

Abbot Silver-Disk Pyrheliometer The silver-disk pyrheliometer was introduced originally as a secondary instrument by the Smithsonian Institution, although it is mainly used now as a transfer standard for calibrating other radiation instruments. More than 100 copies of the silver-disk pyrheliometer have been constructed and distributed, by the Smithsonian Institution, to various countries and institutions throughout the world, the objective being to "diffuse the standard scale of pyrheliometry" as widely as possible. In a resolution of the Subcommission on Actinometry of the World Meteorological Organization meeting in Brussels in 1952, the silver-disk pyrheliometer was recommended as an instrument for the measurement of direct solar radiation.

The design of the instrument can be seen from the diagram of Fig. 3.9

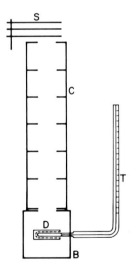

Fig. 3.9 Schematic diagram of the silver-disk pyrheliometer of the Smithsonian Institution. S, shutter; C, collimator tube; D, blackened silver disk; B, copper box; T, thermometer.

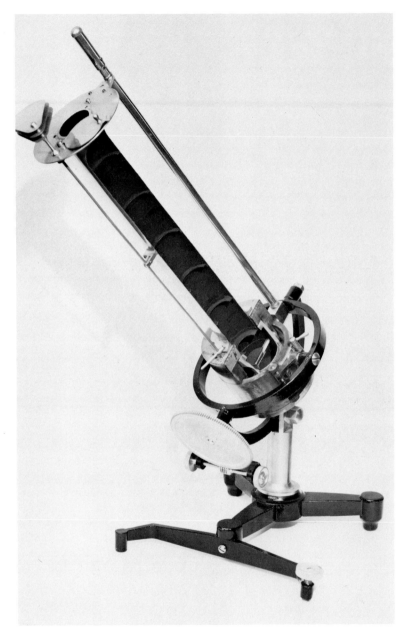

Fig. 3.10 Photo of the silver-disk pyrheliometer with a section removed to reveal
the silver disk and inserted thermometer (courtesy of the Smithsonian Institution).

and the photograph of Fig. 3.10. The sensitive element is a blackened silver disk, D of Fig. 3.9, of dimensions 38 mm in diameter and 7 mm in thickness. The disk has a hole bored radially into its edge, into which is inserted the bulb of a sensitive mercury-in-glass thermometer T. The space surrounding the thermometer bulb is filled with mercury in order to promote a rapid transfer of heat between the silver disk and thermometer. A thin steel lining in the hole prevents amalgamation of the mercury and silver. Finally, the mercury is kept from escaping from the hole by a seal of cord, shellac, and wax surrounding the thermometer stem. The disk is suspended by three fine steel wires inside a copper box B, which is enclosed in a wooden box to protect the instrument from temperature changes of the sur-roundings.

In the original version of the instrument, which was developed during the first years of this century, the sensing element was a blackened disk-shaped copper box filled with mercury. In a later model the mercury in the copper box was replaced by copper filings, but an excessive lag of the in-strument response resulted. The copper box was replaced by a copper disk in 1906, and the final change to the silver disk was introduced about 2 years later. Since 1909 the only change of the instrument has been substi-tution of longer tubes on some models, the objective being to decrease the circumsolar sky radiation received.

The essential operation of the silver-disk pyrheliometer is as follows. Solar radiation enters the instrument through the collimator tube labeled C in Fig. 3.9, is absorbed by the blackened silver disk D, and causes a rise in temperature of the disk. The rate of change of temperature of the disk is monitored by careful readings of the thermometer in a series of 2-min cycles. The field of view of the instrument is limited (by appropriate diaphragms inside the collimator tube) to a circular cone of whole aperture angle of 5.7°. The thermometer stem is bent at a right angle to make the instrument more compact and easier to use. It is graduated in tenths of degree Celsius from -15 to $+50°C$, and the readings are made to hun-dredths of degrees.

A three-wafer shutter of polished metal plates is rotated in and out of the field of view to alternately shade and expose the silver disk to solar radiation in a specified and carefully timed sequence. The requirements for timing the shutter and reading the thermometer are very demanding, even for a trained observer. The procedure originally specified and still normally followed is quoted from Abbot (1911) as follows:

> Having adjusted the instrument to point at the Sun and opened the cover, read the thermometer exactly at 20 seconds after the beginning of the first minute. Read again after 100 seconds, or at the beginning of the third minute,

and immediately after reading open the shutter to expose to the Sun. Note that the instrument is then correctly pointed. After 20 seconds read again. After 100 seconds more (during which the pointing is corrected frequently), or at the beginning of the fifth minute, read again, and immediately close the shutter. After 20 seconds read again. After 100 seconds read again, or at the beginning of the seventh minute, and immediately open the shutter. Continue the readings in the above order, as long as desired. Readings should be made within 1/5 second of the prescribed time. Hold the watch directly opposite the degree to be observed, and close to the thermometer. Read the hundredths of degrees first, and the degree itself afterward.

After the temperature readings obtained during the sequence are corrected for air, stem, and bulb temperatures, the final values of energy flux are obtained by multiplying the corrected temperature difference by a calibration constant for the particular instrument being used. The calibration constant is normally obtained, by the Smithsonian Institution, by a direct comparison of the silver-disk pyrheliometer with a similar working standard which has been referred to the water-flow pyrheliometer as the primary standard. However, care must be exercised in the interpretation of the final energy values obtained with the Smithsonian instruments. The calibration constants furnished with the instruments refer to the Smithsonian 1913 Scale of Radiation, and that scale is known to be too high. To reduce the measurements to the International Pyrheliometric Scale 1956, the values so obtained should be reduced by 2.0%.

Experience with silver-disk pyrheliometers has shown that the timing of the sequence of readings must be very accurate, a consistent error of 1 sec resulting in about a 1% error in the final result (W.M.O., 1965). In an effort to simplify the observational procedure and to reduce the subjectivity of the readings, Foster and Macdonald (1955) devised an automatic shutter operated with a solenoid connected to a synchronous electric clock. By eliminating the timing errors in manually operating the shutter, it was found that the 20-sec readings inserted in the sequence could be eliminated without affecting the accuracy of the results. The revised procedure consists of three temperature readings at 2-min intervals. Reading of the thermometer itself was simplified by installation of a crank-activated pointer, by which the movement of the mercury column is followed until the exact time for the reading. Stopping the movement of the pointer at that instant permits reading the position of the stationary pointer instead of that of the moving mercury column, thereby easing the requirements of reading and increasing the accuracy.

Foster and Macdonald also devised an automatic temperature recorder for the silver-disk pyrheliometer, but its use did not significantly increase the accuracy of the final results.

Various comparisons of the silver-disk pyrheliometer with the double-chamber water-flow pyrheliometer as the primary standard show that the characteristics of the former are extremely stable with time. For instance, Hoover and Froiland (1953) summarize the behavior of one instrument (No. S.I.5$_{bis}$) over a 20-year period in the following tabulation.

Date	No. of values	Mean constant
1932	37	0.3625
1934	42	0.3629
1947	18	0.3626
1952	100	0.3622

They consider the indicated variations to be probably within the error of observation.

Ångström Electrical Compensation Pyrheliometer   The electrical compensation pyrheliometer, developed by Knut Ångström (1893, 1899), was probably the first reliable pyrheliometer and is still one of the most accurate, as well as convenient, instruments for measuring radiant energies. As early as 1905, the International Meteorology Committee meeting in Innsbruck and Oxford recommended that measurements of the total solar radiation be made at central observatories and that "Ångström's compensation pyrheliometer should be used exclusively for these measurements."

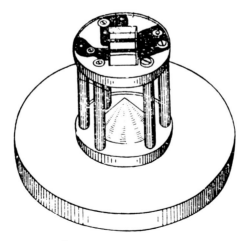

Fig. 3.11 The base of the Ångström electrical compensation pyrheliometer showing the blackened receiving strips (after K. Ångström, 1899).

The principle of operation of the instrument is as follows. Two strips of manganin foil of approximate dimensions $20 \times 2 \times 0.01$ mm are coated on one side with Parsons optical black lacquer (formerly platinum black plus a thin layer of camphor soot was used). The strips are mounted side by side across the opening of the strip holder shown by Fig. 3.11. A thermojunction is attached to the back side of each strip, and electrical leads from the strips and the thermojunctions are attached to external binding posts on the base of the holder. The holder is mounted at the base of a cylindrical metal tube T shown in the photograph of Fig. 3.12, which is itself attached to an azimuth-elevation mechanism by which the tube can be directed at the Sun. A reversible shutter at the front end of the tube permits one of the strips to be shaded from the Sun, while the other is exposed to the direct solar radiation. In operation, the shaded strip is heated by an electric current to the same temperature as that of the exposed strip, at which

Fig. 3.12 A traditional model of the Ångström electrical compensation pyrheliometer (photograph from Science Museum, London).

time the rate of energy absorption by the exposed strip is thermoelec-
trically compensated by the rate of energy supplied electrically to the
shaded strip. Equivalence of temperature of the strips is determined by
the thermojunctions attached to their reverse sides, being connected in
opposition through a sensitive null detector.

Although the Ångström instrument is normally calibrated against a
primary standard, it has the capability in itself of absolute-energy determinations (K. Ångström, 1899). If q is the radiant energy per unit area
and unit time incident on the exposed strip, i is the electric current through
the shaded strip, and b the width, a the absorptance, and r the resistance
per unit length of the strips, then equivalence of energy input to the strips
occurs when:

$$qab = cri^2 \qquad (3.15)$$

where c is a constant which is determined by the units employed. If q is
in cal cm^{-2} min^{-1}, b is in cm, r is Ω cm^{-1}, and i is A, then c = 14.33.

For an absolute measurement of radiation, the instrumental factors
a, b, and r must be determined by some means. For an accuracy of the
instrument to better than $\pm 0.5\%$, each of the factors should be known to
at least $\pm 0.1\%$. Since the determination of a, b, and r to the required accuracy is a difficult procedure, a constant k given by

$$k = cr/ab \qquad (3.16)$$

is normally determined by the manufacturer of the instrument, by reference to a primary standard. Once k is determined for a particular instrument, then the flux of incident radiation is obtained by the simple relation

$$q = ki^2 \qquad (3.17)$$

Reference standards are available at the WMO World Radiation
Center (Davos Observatory, Switzerland) and, by transfer, at the dependent regional centers whose standards are compared, at Davos, at approximately 5-year intervals.

Observations with the electrical compensation pyrheliometer consist
of a series of measurements (normally at least ten for highest accuracy)
in which each strip is alternately exposed and shaded. With precision electrical measurement equipment and good observing conditions, reproduction of the International Pyrheliometric Scale (IPS 1956) to within $\pm 0.2\%$
is possible, which infers an accuracy of 1% or better (CSAGI, 1958).

The high degree of stability of Ångström-type pyrheliometers is
demonstrated by the following selected results obtained during the 1970
Davos (and Locarno) International Pyrheliometric Comparison (Table

3.5). All instruments were referred to the Davos (WMO) standard Ångström pyrheliometer No. 210. With the exception of Sweden (and Argentina and Australia, based upon Sweden), the agreement of the regional and national standards, mainly carrying the 1964 calibrations, which were determined similarly, was generally within $\pm0.5\%$; one-third of the pyrheliometers agreed with the WMO reference to within $\pm0.25\%$. All comparisons were centralized to the Davos digital sampling and read-out system, the individual compensation currents (after manual balance of the strips) being read nearly simultaneously as voltages across a series of standard resistors.

TABLE 3.5

Stability of Ångström-Type Pyrheliometers

Country	Inst. No.	Obs.	A_x/A_{210}
A. Regional			
Argentina	553	200	0.984
Australia	578	83	1.020
Belgium	7190	202	0.995 ⎫ 1.000
	21	199	1.004 ⎭
Canada	9001	110	0.996
Congo	119	124	1.009 ⎫ 1.005
	10	71	1.001 ⎭
Great Britain	583	110	0.994
India	508	175	0.997 ⎫ 0.994
	8418	24	0.991 ⎭
Japan	588	182	1.021 ⎫ 1.006
	10316	24	0.992 ⎭
South Africa	542	144	1.001
Sudan	561	179	1.001
Sweden	158	164	1.011 ⎫ 1.014
	171	198	1.016 ⎭
Switzerland	525	205	0.997
Tunisia	205	102	0.998 ⎫ 1.003
	9002	92	1.008 ⎭
USA	2272	202	0.995 ⎫ 0.997
	7644	14	0.999 ⎭
USSR	212	103	1.003
B. National			
France	24	41	1.000
Nigeria	576	194	0.996
Norway	204	206	0.990 ⎫ 0.996
	507	200	1.002 ⎭

With regard to instrument construction, most of the earlier models suffered from the so-called "edge effect" caused by the exposed strip being shaded at the ends by the innermost diaphragm, while the shaded and electrically heated strip has the same temperature over its whole length. It is estimated that this introduced an error of the order of -2.0% (A. K. Ångström, 1958) but, because of construction differences in the individual instruments, the corresponding correction was never applied in routine meteorological practices. In the recent designs, originated by the Swedish Meteorological and Hydrological Institute and the Eppley Laboratory, this effect is largely eliminated through arranging for the screening to occur exactly at the terminations of the strip. Other possible sources of error arise from the inability to heat the shaded strip in the same manner as the exposed one, and from differences in the thermal conductance of the strip blackening. Finally, small optical effects may be introduced within the diaphragm system of the pyrheliometer tube and behind the strips. In the Eppley model, a blackened conical shield is placed behind the strips to minimize such reflections and emissions, and the element is of permanent soldered construction for improved instrument stability. Both modern Swedish and Eppley models incorporate microswitches, which orient the heating current simultaneously with strip selection.

The IGY Instruction Manual (CSAGI, 1958) specifies that for instruments calibrated according to the (uncorrected) Ångström Scale befor 1 January 1957, measured radiation values should be multiplied by the factor 1.015 to convert them to the IPS. Pyrheliometers which have been standardized since 1957 have this correction included in the instrument constant.

An important aspect of pyrheliometer design is the choice of the instrument acceptance aperture which, unfortunately, has varied greatly in the past. The aperture (rectangular) angles of the older models are about $6° \times 24°$. In 1935 the Smithsonian Institution introduced a model with a reduced aperture of $3° \times 6°$, mainly for use in its network of solar-constant observing stations. The corresponding values for the newer designs are intermediate, e.g., the Eppley aperture is $4.5° \times 10.5°$ approximately; an effort is being made to standardize two current models in this respect. However, since the strip screened from the Sun also views a portion of the circumsolar sky (but not the same portion as that viewed by the exposed strip), there is a measure of optical compensation introduced. This has the effect of modifying the geometrical apertures of the tube system to the effective pyrheliometric aperture of A. K. Ångström and Drummond (A. K. Ångström and Rodhe, 1966); in the earlier Swedish instruments this is equivalent to a circle of about $8°$ whole angle and, in the later ones, to about $5°$.

Many types of auxiliary instrumentation are available for precise regulation of the compensating electrical current. For example, the Eppley Ångström control unit incorporates a precision multipoint current potentiometer for checking the milliammeter to within 0.05% during field operation. Also, digital voltmeters, with or without printers, may be connected across standard resistors in the current circuit. In such systems, the null detector preferred is one of a suitable sensitive electronic type. Marsh (1965) has recently introduced a servo-operated electronic system by which the current to the heated strip is controlled automatically, thereby rendering the operation more convenient and possibly reducing personal errors in the measurements.

3.2.4 Operational Pyrheliometers

Eppley Normal Incidence Pyrheliometer   There has been built up over the years, in the United States, a confusion of terms with regard to the instruments made by the Eppley Laboratory. For a long time the instrument for measuring radiation from the whole hemisphere (pyranometer in the terminology used here) was called a pyrheliometer, by both the manufacturer and the users of the instruments. However, on introduction of the instrument for measuring direct solar radiation only (pyrheliometer) some differentiation in nomenclature was required, so for some time the term "180° pyrheliometer" was applied to the pyranometer and the direct radiation instrument was called a "normal incidence pyrheliometer." More recently the hemispherical instrument has been termed a pyranometer, in accord with the recommendations of the World Meteorological Organization (1965), while "normal incidence" is still associated with the pyrheliometer.

The configuration of the Eppley normal incidence pyrheliometer can be seen from Fig. 3.13.

The sensitive element of the pyrheliometer, in its present form, is a fast-response wire-wound plated (copper-constantan) multijunction thermopile with a 9-mm-diameter receiver coated with Parsons optical black lacquer. A temperature-compensating circuit to stabilize sensor response is incorporated in the heat sink of the thermopile. The unit is mounted at the base of a double-wall brass tube which is chromed externally and blackened internally. A series of diaphragms limits the aperture to a circular cone of full angle 5.7°. The innermost diaphragm screens a 1-mm annulus of the receiver at its edge; this somewhat reduces the sensitivity variation of the thermopile with deviation from perfect solar tracking. The arrangement of an air-insulated double-walled cover significantly increases

Fig. 3.13   The Eppley normal incidence pyrheliometer (courtesy of the Eppley Laboratories).

instrument stability, especially during windy weather. The interior of the system if filled with dry air at atmospheric pressure, and the viewing end is sealed by a removable insert carrying a crystal quartz window of 1 mm thickness. Two flanges, one at each end of the pyrheliometer tube, are provided with sighting devices to allow the instrument to be oriented toward the Sun. A manually rotatable disk, which can accommodate three filters (such as Schott OG1, RG2, and RG8) and leave one aperture for total spectrum measurements, is provided.

The general instrument characteristics are given in the following tabulation.

Sensitivity	5–6 mV/cal cm^{-2} min^{-1}
Impedance	200 Ω (approximately)
Temperature dependence	± 1% over range −30 to +40°C
Linearity of response	linear up to 4 cal cm^{-2} min^{-1}
Response time	1 sec (1/e signal) or 4 sec (maximum)

In earlier models of this pyrheliometer, an 8-junction copper-constantan thermopile, with a 5-mm-diameter receiver coated with lampblack, was first employed. The sensitivity was about 2 mV/cal cm^{-2} min^{-1}, impedance 6 Ω and response time (maximum) 6 sec. This unit was replaced, in 1958, by a temperature-compensated, 15-junction bismuth-silver thermopile with a receiver coated with Parsons black lacquer. The sensitivity was about 3 mV/cal cm^{-2} min^{-1}, impedance 450 Ω, and response time (maximum) 20 sec. The current model was introduced in 1965.

The signal from the Eppley pyrheliometer is suitable for remote recording, and the case is weatherproofed for continuous operation in an exposed location. In practice, the instrument is normally attached to an electrically driven equatorial mount for solar tracking, and the output signal is either recorded directly on a strip-chart recorder or integrated over an appropriate time period and registered. The instrument has been found to be very stable with time, and by occasional restandardization it is suitable for use as a secondary standard for calibrating other pyrheliometers. It was recommended as one of the instruments for solar radiation measurements during the 1957–1958 International Geophysical Year.

Linke–Feussner Pyrheliometer (Actinometer)   Ladislaus Gorczynski (1924) built the first pyrheliometer using the Moll thermopile, for making measurements of solar radiation in the Sahara Desert. The pyrheliometers (pyrheliographs) of the original Gorczynski design, two with spectral filters and one unfiltered, were in daily use at the Kew Observatory in 1950 (Stagg, 1950). The unfiltered one had been used continuously since 1932. It incorporated the large-surface 80-junction Moll thermopile, with appropriate shading diaphragms, and an equatorial mount driven by a pendulum clock. Later versions of instruments based on the Moll thermopile have culminated in the Linke–Feussner pyrheliometer.

The Linke–Feussner pyrheliometer uses a specially designed Moll thermopile consisting of 40 manganin-constantan thermocouples arranged in a circle of 1 cm diameter. The thermocouples are in two equal sectional arrays, which are connected in opposition. One section is exposed to the radiation being measured and the other is shaded. Thus, the sections tend to compensate each other for short-period temperature fluctuations of the environment and for thermal effects caused by quasiadiabatic pressure changes near the thermopile surface which occur in fluctuating air currents. The sensitivity of the thermopile* is about 11 mV/cal cm^{-2} min^{-1}, a value which is somewhat higher than that usually available in pyrheliometers. Because of this high sensitivity, the instrument can be used advantageously for measuring radiation from the sunlit sky, as well as for direct solar radiation. Impedance of the thermopile is about 65 Ω. The small mass of the thermocouple elements and the high heat conductivity of their mounting posts makes the Moll thermopile respond rapidly to changes of radiation. The instrument as it is presently configured shows 99% response in 8–10 sec. The acceptance aperture is a cone of 10.2° total angle, and the diameter of the opening of the entrance diaphragm is 12.6 mm.

The main body of the instrument, shown by the photograph of Fig. 3.14, is made up of six massive copper rings which are contoured on the inside to produce a set of radiation diaphragms for decreasing internal reflections, for defining the acceptance angle of the instrument, and for limiting turbulent air currents inside the instrument. This turbulence limiting feature, combined with the compensating thermopile, makes the instrument suitable for use in windy conditions. No mechanism is provided for compensating for the change of signal from the thermopile with the instrument temperature. The temperature dependence is expressed by the factor $[1 + \alpha(T - 20°C)]$, where T is temperature, α has a value of approximately 0.002, and the calibration temperature is 20°C. The effects of short-period changes of environment are minimized by the body being of massive copper rings for providing a high heat capacity and for equalization of temperatures throughout the instrument interior. Heat transfers with the environment are minimized by a layer of felt insulation around the copper body and by a shield over the front surface. A thermometer embedded in the mass of copper permits a determination of instrument temperature, from which temperature corrections can be made.

Since the thermopile is sensitive to radiation out to at least 40 μm, it is necessary for solar radiation measurements to limit the spectral range of radiation which enters the aperture. This is done by the use of radiation filters which are installed in the filter holder mounted on the front end of the instrument body. Provision is made for inserting five different filters in the filter holder. For the exclusion of the long-wave terrestrial radiation a filter of ultrasil (quartz) is provided, and the absorption filters OG1 (transparent from 0.525 to 2.80 μm) and RG8 (0.70 to 2.80 μm) are normally used in three of the filter mounts. The fifth mount is fitted with a double-

---

* Performance specifications are partially from Kipp and Zonen Bulletin SOL 66.

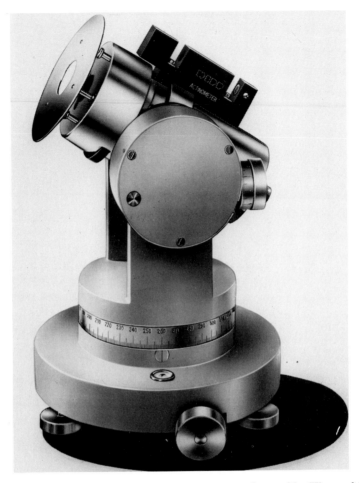

Fig. 3.14 The Linke–Feussner pyrheliometer as manufactured by Kipp and Zonen, Delft, Holland (courtesy of the manufacturer).

walled opaque disk for use in zeroing the instrument. The effects of the very troublesome increase of temperature of the color filters by the absorption of solar radiation (Drummond and Roche, 1965) is minimized, but not eliminated, by enclosing the filter assembly inside another massive copper ring which is in good thermal contact with the copper body of the instrument.

As normally supplied by the manufacturer (Kipp and Zonen, Delft, Holland), the instrument is installed on a manually operated azimuth-elevation mount by which it can be oriented in any direction of the hemisphere. For use in monitoring direct solar radiation it can be conveniently attached to an appropriate equatorial mount.

Michelson Bimetallic Pyrheliometer   The Michelson pyrheliometer was originally constructed in Moscow at about the same time as the Abbot silver-disk pyrheliometer (Michelson, 1908). It is in reality a bimetallic thermometer, a schematic diagram of which is shown in Fig. 3.15. More recent constructions with technical improvements which are currently available, are those modified by Beuttner (1930) and Marten (1931).

In principle, the deflection of a delicate fiber I attached to the free end of a fine bimetallic (usually constantan-invar) blackened lamina S, irradiated by the solar beam A, is observed through a relatively low-power microscope M. The aperture of the instrument has angles of approximately 5° and 13°. If handled carefully, the instrument is extremely reliable; it is of reasonably rapid response (20–30 sec for a 100% intensity change). There are, however, some important precautions which should not be neglected. The zero position of the lamina, which is observed without solar radiation, is, for the original and Marten models, dependent on the instrument temperature and therefore shows a considerable shift. For determining and eliminating this shift it is necessary to control the zero position after each radiation measurement; a series of successive measurements must therefore consist of alternate observations with and without radiation. The time interval between the individual measurements must be constant, e.g., 20 or 30 sec, according to the interval used for the calibration of the instrument. In addition, it is recommended that the actinometer be exposed to the Sun for 10 min before the start of a series and to irradiate the lamina for 1 min before beginning the observations. This procedure minimizes the zero shift. These complications are avoided in the Beuttner construction, where the zero position is largely temperature compensated by incorporation of a second similar but opposed and radiation-screened bimetallic lamina; the zero position need be observed only at the beginning and at the end of every series of measurements, and the intervals between the individual measurements need not be kept absolutely constant.

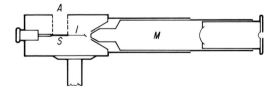

Fig. 3.15 Simplified diagram of the Michelson bimetallic pyrheliometer. A, aperture; S, blackened lamina; I, fiber; M, microscope.

The calibration factor must be determined by comparison with another well-calibrated pyrheliometer; it is expressed as the caloric value of one scale unit. In principle, the calibration factor of this bimetallic instrument has a dependence on the instrument temperature which has to be determined experimentally. This temperature dependence has to be taken into consideration for routine measurements as well as for calibration.

On account of its portability, the bimetallic pyrheliometer is especially suitable for use as an instrument for daily routine measurements and also as a travelling substandard in a radiation network. It can be equipped with a quartz window and interchangeable filters. Use of the former (with appropriate change of calibration factor) is advantageous in windy conditions; with the recommended filters, measurements of atmospheric turbidity are readily obtainable. However, the fragile nature of the instrument and the need for frequent recalibration, for standardizing purposes, has resulted in its limited popularity.

Savinov–Yanishevsky Pyrheliometer (Actinometer)   This is the main operational pyrheliometer for direct solar radiation measurements in the U.S.S.R. It is used as a relative instrument, and therefore requires calibration against an absolute standard. The primary standard in the U.S.S.R. is the Ångström electrical compensation pyrheliometer.

The sensor of the Savinov–Yanishevsky pyrheliometer is a thin blackened silver disk, 11 mm in diameter, from which the center section of 3.5 mm diameter has been removed (Kondratyev, 1965; Yanishevsky, 1957). To the back of the disk are cemented the hot junctions of 36 pairs of

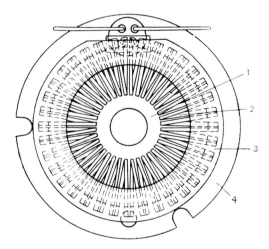

Fig. 3.16   Diagram showing the thermocouple arrangement (the "thermostar") of the Savinov–Yanishevsky pyrheliometer (after Yanishevsky, 1957).

thermocouples made from $6.0 \times 0.3 \times 0.04$ mm strips of manganin and constantan. The cold junctions are cemented to a heavy copper ring, which is in good thermal contact with the main body of the instrument. The radial orientation of the thermocouples has given rise to the designation "thermostar" (*termo-zvezdockkoi*) for the sensor assembly (see Fig. 3.16).

The thermostar is mounted at the bottom of a collimator tube, the aperture of which is a $5.0°$ circular cone. Conductors from the thermostar are led out from the base of the instrument to an indicator or recorder, the main recorders in use being of the galvanometric or balancing potentiometric type. The internal resistance of the thermostar is 15–20 Ω, and the output signal is 4–7 mV for an incident flux of 1.0 cal cm^{-2} min^{-1}. The response of the instrument is approximately proportional to the flux of incident solar radiation. The instrument is not temperature compensated; the temperature dependence is about $-0.1\%/°C$ rise in ambient temperature. It is used for continuous measurements of the direct solar radiation by being installed on a clock-driven equatorial mount.

Pyrheliometer of the Japanese Meteorological Agency   A recent development is an experimental model of a pyrheliometer designed and built by the Weather Instruments Plant of the Japan Meteorological Agency (Fujimoto, 1965). It is based on a Moll-type thermopile consisting of 8 pairs of copper-constantan junctions. The thermopile is mounted at the base of a brass tube of 3 cm diameter and 13 cm length which is blackened inside and chrome-plated outside. Diaphragms in the tube limit the aperture to a $6°$ cone. Calibration with a silver-disk pyrheliometer shows an instrumental constant of 1.51 mV/cal cm^{-2} min^{-1} with a mean error of ± 0.02 mV. According to the authors, it is subject to about the same temperature errors as are other instruments of similar design, and no temperature compensation is provided. Further details on the operating characteristics have not yet appeared.

3.3   COLLIMATOR TUBES FOR PYRHELIOMETERS

One of the major problems in measuring the direct solar radiation and in establishing a firm scale of radiant energy has been the divergence in results from pyrheliometers of different construction because of differences in the amount of circumsolar sky seen by a pyrheliometer detector. It is impossible to eliminate this sky radiation completely from the measurements because of (a) the finite dimensions of the components, (b) the practical difficulty of exact orientation of the instrument, and (c) the inability to define the solar disk precisely. The intensity of radiation from the circumsolar sky is a function of solar elevation, station altitude, and

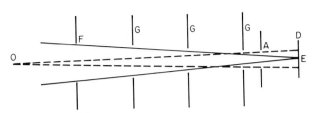

Fig. 3.17 Diagram showing the optical configuration of collimator tubes of pyrheliometers. E, point on detector D; A, aperture stop; F, field stop; G, glare stops; O, point on optical axis.

atmospheric conditions, so the introduction of corrections for the effect is an involved and somewhat unreliable procedure with our present knowledge.

Computations have shown (Pasteils, 1959) that under average clear-sky conditions at sea level, the circumsolar sky radiation may introduce a 2–3% error in measurements of direct solar radiation made with the Ångström electrical compensation pyrheliometer or the Smithsonian silver-disk pyrheliometer. Other instruments may have different characteristics, depending on their individual designs.

The optical configuration of collimator tubes is usually made as simple as possible in order to eliminate the undesirable effects of lenses, mirrors, or other optical components. In the majority of instruments the solid angle viewed is conical in form, although there are notable exceptions. For instance, the field of the Ångström pyrheliometer is rectangular, the angular dimensions of which vary considerably among different models of the instrument.

For purposes of illustration of the aperture and field of view, we assume symmetry around the optical axis. The aperture stop A of Fig. 3.17 limits the cross-sectional area of the beam which illuminates the detector D from points on the optical axis, such as point 0, since all other elements are large enough to accept a larger cone. The field stop F limits the field of view— the solid angle which can be seen by the detector element E that is on the optical axis of the system. Glare stops G are only for decreasing stray light in the system.

From point E, the field of view for this case is circular in shape, but for off-axis points of the detector a certain amount of vignetting occurs.

The International Radiation Commission (1956) has attempted to bring a measure of standardization into the geometry of pyrheliometer tubes. If R is the radius of the field stop, r the radius of the aperture (receiver), and l the distance between field stop and receiver, .we define the opening angle Z_o, slope angle Z_p, and limit angle Z by the respective relations

$$\tan Z_o = R/l \quad \tan Z_p = (R - r)/l \quad \tan Z = (R + r)/l \quad (3.18)$$

The recommendation of the International Radiation Commission is that pyrheliometers should have a slope angle not less than 1° or greater than 2° ($1° \leq Z_p \leq 2°$) and a ratio of distance between receiver and field stop to radius of receiver of at least 15 ($1/r \geq 15$). These conditions imply that the opening angle $Z_o \leq 4°$. No specifications for the limit angle Z were given.

Data for the Smithsonian silver-disk pyrheliometer and the Eppley pyrheliometer are given in the following tabulation (Drummond, 1961).

	Smithsonian	Eppley
Z_o	2.9°	2.9°
Z_p	0.8°	2.1°
Z	4.9°	3.6°

In both cases, Z_p is outside the recommended range, being larger than recommended for the Eppley instrument and smaller for the Smithsonian instrument, while Z_o is within the recommended range for both.

The above discussion is based on the assumption of circular optics, but a rectangular field stop is combined with a rectangular receiver in the Ångström pyrheliometer. The Ångström instrument has a rectangular field with slope angles of approximately 3° × 12° in the original version, and about 1.75° × 7° in a more recent model (Lindholm, 1958). In both cases the slope angles considerably exceed those recommended by the International Radiation Commission. Because of the wide field of view, measurements with the Ångström instruments are particularly susceptible to errors introduced by radiation from the circumsolar sky. Pastiels (1959) has estimated that because of its larger field, measurements with the Ångström pyrheliometer have an error of 0.5% greater than that of the Smithsonian instrument under average clear-sky conditions.

3.4   CALIBRATION OF PYRHELIOMETERS

Pyrheliometers which are normally used as operational instruments, including the Eppley, Linke-Feussner, Savinov–Yanishevsky, and other recording types, are not designed to be absolute instruments, so they require calibration against standard instruments in order that their indications may be interpreted in terms of absolute energy. The main standard pyrheliometers which have been developed are the Ångström electrical compensation pyrheliometer and the Smithsonian water-flow pyrheliometer, both of which were discussed above. The water-flow pyrheliometer is somewhat complicated and difficult to use, and it has not been employed by the Smithsonian Institution for absolute-energy determinations since the termination of its solar constant observing program in the early 1960's. This leaves the Ångström pyrheliometer as the primary standard to which other solar radiation instruments are referenced.

Although the Ångström instrument is designed to yield absolute-energy determinations from first principles, there are difficulties in achieving this objective. In practice, it is almost invariably operated in the relative manner, requiring calibration against an absolute standard. The generally recognized solar radiation reference is a group of Ångström pyrheliometers maintained at the World Radiation Center at the Physical Meteorological Observatory, Davos, Switzerland. In 1959, as part of the First International Pyrheliometric Comparison, primary standard Angström pyrheliometers maintained at Stockholm, Sweden were compared with the Davos standards, in an effort to perpetuate the original Uppsala Ångström (No. 70) standard. Since then, at the subsequent Davos international comparisons in 1964 and 1970, this link with the Swedish references has continued. Because of the former use of various scales of radiation, care must be taken to reference calibration factors to the proper scale. The presently accepted scale is the International Pyrheliometric Scale 1956 (IPS).

Working standard pyrheliometers of the Ångström type (referenced to Davos) are maintained at the regional and national centers designated by the World Meteorological Organization for the purpose of reproducing the IPS. To a limited extent, the Abbot silver-disk pyrheliometer is still employed at such centers. The secondary transfer instruments in current use are other Ångström pyrheliometers and instruments of the Eppley, Linke–Feussner, and Michelson types.

Once a good secondary instrument is properly calibrated, there is little difficulty in using it to transfer this calibration to operational instruments. The comparison should be carried out when there is as little haze, smoke, or other obscuring materials as possible in the atmosphere, and under conditions in which no clouds are within 20° of the Sun. The two instruments should be installed in close proximity to each other on mounts by which they can easily be oriented to the Sun. A sufficient number (at least ten is suggested) of comparisons should be made between the secondary standard and the operational instrument to assure a representative constant for the latter. If $\bar{F}$ is the average flux of solar radiation determined from n readings of the secondary standard and $\bar{r}$ is the average response (in millivolts or other units) determined by an equivalent number of readings of the operational instrument, then the required constant c is given by

$$c = \bar{F}/\bar{r}$$

In cases for which the calibration constant is not independent of incident flux, the above procedure should be repeated at several different solar elevations, but for most of the commercially available pyrheliometers the single value of c is adequate. The frequency with which a given instrument should be calibrated is dependent on many factors, but one in continuous operation should be recalibrated several times per year.

Documentation of the calibration is very important and should not be neglected. A calibration certificate containing all pertinent data (instrument numbers, date, location, observers, radiation values, etc.) should be completed and certified by the appropriate authority immediately on termination of the calibration, and the certificate should be placed on permanent file in a secure and convenient location for possible future reference. A series of calibration certificates can be used to assess the behavior of the instrument over a period of time, and to yield a valid basis for correcting the data should such prove necessary.

REFERENCES

Abbot, C. G. (1911). The silver disk pyrheliometer. *Smithson. Misc. Coll.* **56** (19), 10 p.

Abbot, C. G., and Aldrich, L. B. (1932). An improved waterflow pyrheliometer and the standard scale of solar radiation. *Smithson. Misc. Coll.* **87** (15), 8 p.

Abbot, C. G., Fowle, F. E., and Aldrich, L. B. (1922). New evidence on the intensity of solar radiation outside the atmosphere. *Smithson. Inst. Annu. Astrophys. Obs., App. 1*, **4**, 323–366.

Angstrom, A. K. (1929). On the atmospheric transmission of Sun radiation and on dust in the air. *Geogr. Ann.* **11**, 156–166.

Angstrom, A. K. (1930). On the atmospheric transmission of Sun radiation II. *Geogr. Ann.* **12**, 130–159.

Angstrom, A. K. (1958). On pyrheliometric measurements. *Tellus* **10**, 342–354.

Angstrom, A. K. (1961). Techniques of determining the turbidity of the atmosphere. *Tellus* **13**, 214–223.

Ångström, A. K. (1970). Apparent solar constant variations and their relation to the variability of atmospheric transmission. *Tellus* **22**, 205–218.

Angstrom, A. K., and Drummond, A. J. (1961). Basic concepts concerning cutoff glass filters used in radiation measurements. *J. Meteorol.* **18**, 360–367.

Ångström, A. K. and Rodhe, B. (1966). Pyrheliometric measurements with special regard to the circumsolar sky radiation. *Tellus* **18**, 25–33.

Ångström, K. (1893). Eine elegtrische Kompensationsmethode zur quantitativen Bestmmung strahlender Warme. *Nova Acta Reg. Soc. Sc. Upsal., Ser. III*, No. 16 [Transl. *Phys. Rev.*, **1** (1894)].

Angstrom, K. (1899). The absolute determination of the radiation of heat with the electric compensation pyrheliometer, with examples of the application of this instrument. *Astrophys. J.* **9**, 332–346.

Anon. (1885). Sixth annual exhibition of instruments. *Quart. J. Roy. Meteorol. Soc.* **11**, 242–250.

Aufm Kampe, H. J. (1950). Visibility and liquid water content in clouds and in the free atmosphere. *J. Meteorol.* **7**, 54–57.

Beuttner, K. (1930). *Beitr. Physik. Freien Atm.* **10**, 97.

Bilton, T., Flowers, E. C., McCormick, R. A., Kurfis, K. R. (1973). Atmospheric turbidity with the dual–wavelength sunphotometer. Background Papers for the NOAA Solar Energy Data Workshop, Silver Spring, Maryland, November 29–30, 1973 (mimeographed).

Coulson, K. L., and Gray, E. L. (1965). Unpublished computations.

Covert, R. N. (1925). Meteorological instruments and apparatus employed in the U. S. Weather Bureau. *J. Opt. Soc. Amer. Rev. Sci. Inst.*, **10**, 299–425.

CSAGI (1958). Radiation instruments and measurements, Part 4, "IGY Instruction" Manual," pp. 371–466. Pergamon, Oxford.

Deirmendjian, D. (1969). "Electromagnetic Scattering on Spherical Polydispersions." Amer. Elsevier, New York.

Dorno, C. (1922). Progress in radiation measurements. *Mon. Weather. Rev.* **50**, 515–521.

Drummond, A. J. (1961). Current developments in pyrheliometric techniques. *Sol. Energy* **5**, 19–23.

Drummond, A. J., and Roche, J. I. (1965). Corrections to be applied to measurements made with Eppley (and other) spectral radiometers when used with Schott colored glass filters. *J. Appl. Meteorol.* **4**, 741–744.

Eddy, J. A. (1961). The stratospheric solar aureole. Doctoral Dissertation, Univ. of Colorado, Boulder, Colorado.

Feussner, K., and Dubois, P. (1930). Trubungsfaktor, precipitable water. *Beitr. Geophys. Gerlands* **27**, 132.

Foster, N. B., and Macdonald, T. H. (1955). Silver disk pyrheliometry simplified. *Mon. Weather. Rev.* **83**(2), 33–37.

Fraser, R. S. (1959). Scattering properties of atmospheric aerosols. Sci. Rept. No. 2, Contr. AF19 (604) 2429, Univ. of Calif., Los Angeles, California.

Fujimoto, F. (1965). Actinograph designed by Japan Meteorological Agency. *J. Meteorol.*

Res. Tokyo **17**(10), 610–614 (in Japanese). *Abstr. Meteorol. Geoastro. Abstr.* **18**(6), 1341 (1967).

Glazebrook, Sir Richard (1923). "A Dictionary of Applied Physics," Vol. 3, pp. 699–719.

Gorczynski, L. (1924). On a simple method of recording the total and partial intensities of solar radiation. *Mon. Weather. Rev.* **52**, 299–301.

Hand, I. F. (1937). Review of United States Weather Bureau solar radiation investigations. *Mon. Weather. Rev.* **65**, 415–441.

Hoover, W. H., and Froiland, A. G. (1953). Silver-disk pyrheliometry. *Smithson. Misc. Coll.* **122**(5), 10 p.

Kondratyev, K. Ya. (1965). "Actonometry," Hydrometeorological Publishing House, Leningrad (in Russian); translation NASA TT F-9712 (Nov. 1965).

Lindholm, F. (1958). On the Angstrom absolute pyrheliometric scale. *Tellus* **10**, 249–255.

Linke, F. (1922). *Beitr. Physik Freien Atm.* **10**, 91.

Linke, F. (1929). *Beitr. Physik Freien Atm.* **15**, 176.

Marsh, V. (1965). A system for automatic operation of the Angstrom pyrheliometer. *Arch. Meteorol., Geophys. Bioklimatol., Ser. B* **13**, 71–75.

Marten, W. (1931). *Gerlands Beitr. Geophys.* **32**, 69.

Michelson, W. A. (1908). Ein neues aktinometer. *Meteorol. Z.* **25**, 246.

Pastiels, R. (1959). Contribution to the study of the problem of actinometric methods. *Publ. Inst. Roy. Meteorol. Belg. Ser. A* (11).

Schuepp, W. (1949). Die bestimmung der komponenten der atmospharischen trubung aus aktinometer mussungen. *Arch. Meteorol.*

Shulgin, W. M. (1927). Improved water–flow pyrheliometer. *Mon. Weather. Rev.* **55**(8), 361–362.

Stagg, J. M. (1950). Solar radiation at Kew Observatory. Geophys. Memoirs No. 86, Meteorological Office, London. 87 pp.

Volz, F. (1959). Photometer mit selen–photoelement zur spektralen messung der sonnenstrahlung und zur bestimmung der wallenlangeabbhangigkeit der dunsttrubung. *Arch. Meteorol., Geophys. Bioklimatol.* **10**, 100–131.

Waldram, J. M. (1945). Measurements of the photometric properties of the upper atmosphere. *Trans. Illum. Eng. Soc.* **10**, 147–187.

Whipple, R. S. (1915). Instruments for the measurement of solar radiation. *Trans. Opt. Soc., London* **15**, 106–182.

World Meteorological Organization (1965). Measurement of radiation and sunshine. *In* "Guide to Meteorological Instrument and Observing Practices," 2nd ed., W.M.O. No. 8, TP3 (Loose Leaf, 1961).

Yanishevsky, Yu. D. (1957). "Actinometric Instruments and Methods of Observation." Hydrometeorological Publishing House, Leningrad U.S.S.R. (in Russian).

CHAPTER FOUR

Solar Radiation: Diffuse Component

Since Rayleigh's famous explanation, in 1871, of the blue color of the sky and of the main features of the polarization field of the diffuse skylight, the scattering problem has engendered a great deal of interest. In order to understand the physical principles involved it is well to review the main features of atmospheric scattering.

A train of electromagnetic waves can be represented as a series of vibrations of the electric (or magnetic) vector in the plane normal to the direction of propagation of the waves. The vibration can further be resolved into components in two mutually orthogonal directions in the tranverse plane, and the radiation can be completely defined by means of the amplitudes of the components together with the phase difference between them. There is no assurance that, in actual radiation, the vibrations are at all regular; there may be incessant time variations of amplitudes and phases. Our sensing devices—the eye, photosensitive elements, etc.—are so crude that they do not respond to the individual vibrations but integrate the energy of the vibrations over a time corresponding to a large number of vibrations. Then we measure an intensity or flux of energy which is proportional to the time average of the square of the individual amplitudes, which tells us little about the individual vibrations themselves. We do know, however, that there must be something regular about the vibrations, because light does exhibit polarization. In the most general type of polarization (elliptic polarization) the end of the electric vector describes an ellipse of some type, so the two orthogonal components must have some constant relationship which holds throughout all the variations.

It can be determined by proper instrumentation that the orientation of the ellipse remains fixed, which means that the phase difference between the components must be constant. The ellipticity of the light also appears to be a fixed quantity, which would mean that the ratio of the orthogonal components is constant. This is not necessary for the individual vibrations, however, as the constancy of phase difference is. As pointed out above, we cannot measure the individual amplitudes, but only a quantity which is proportional to the time average of the square of the amplitudes, i.e., the intensity. It is sufficient for the ratio of the intensities of the components to remain constant.

Sir George Stokes showed, in 1852, that a radiation field in an arbitrary state of polarization can be represented completely by a set of four parameters—the so-called Stokes polarization parameters. They are particularly well suited to analytical work because they all have the dimensions of intensity. In order to interpret the Stokes parameters physically, we denote two orthogonal directions in the plane normal to the direction of propagation of a pencil of radiation by e and r, and let I_e and I_r represent the intensity components along the e and r directions, respectively. Let us now make the following definitions:

$$Q = I_e - I_r \tag{4.1}$$

$$U = 2(I_e I_r)^{1/2} \cos \delta \tag{4.2}$$

$$V = 2(I_e I_r)^{1/2} \sin \delta \tag{4.3}$$

δ is the phase angle between the vibrations along e and r. The quantities I, Q, U, and V are the Stokes parameters for polarized light. By their use we can describe a radiation field completely, regardless of its state of polarization. The total intensity of the light, and therefore its total energy content, is given by

$$I = I_e + I_r \tag{4.4}$$

The characteristics of the polarization field and its measurement are the subjects of Chapter 7; here we are concerned only with the intensity of diffuse skylight and the flux of radiant energy contributed by this skylight.

The energy which constitutes the diffuse radiation from the sunlit sky is scattered from the incident solar beam by scattering on particles of some type which are either suspended in the atmosphere or are part of the atmosphere itself. The two general types of scattering particles which have received the most emphasis in atmospheric investigations are those for which the physical dimensions are small compared to the wavelength of the radiation (Rayleigh scatterers), and those which are of a size comparable to or larger than the wavelength (Mie scatterers). The gaseous molecules of the air, principally oxygen and nitrogen, are by far the most numerous Rayleigh scatterers, and they dominate the scattering pattern under cases of a very clear and haze-free atmosphere. For a turbid atmosphere, however, the aerosol particles scatter so strongly that Mie scattering is of at least equal importance with Rayleigh scattering in blue and ultraviolet wavelengths; in the longer visible and infrared regions the scattering process in the usual slightly turbid atmosphere is dominated by Mie scattering, and for a heavily polluted or cloudy atmosphere, Mie scattering is dominant in all wavelengths.

4.1 SKYLIGHT INTENSITY AND DISTRIBUTION

4.1.1 Case of a Clear Atmosphere

The intensity and flux of radiation for a clear sunlit sky have been investigated theoretically by Chandrasekhar and Elbert (1954), Deirmendjian and Sekera (1954), Sekera (1956), Coulson (1959), deBary (1964), Dave (1964), and many others. Coulson, Dave, and Sekera (1960) have developed an extensive set of tables of the intensity and polarization fields of radiation which is directed downward at the base of the atmosphere and that which is directed outward to space from the top of the atmosphere.

The intensity field of skylight in the clear atmosphere, as computed by Coulson *et al.* (1960), is shown for three different wavelengths in the principal plane, for the case of a sea-level location and no surface reflection, by the solid curves of Fig. 4.1. (Although the ordinate is in relative units, it can easily be changed to absolute-energy units by multiplying by F_0/π, where F_0 is the energy of solar radiation, expressed in the desired units, which is incident at the top of the atmsophere.) Strong horizon brightening is a characteristic feature of skylight at the longer wavelengths, whereas attenuation of the scattered radiation through the long optical pathlengths near the horizon causes horizon darkening at the shorter wavelengths. The increase of scattering efficiency with decreasing wavelength, according to the λ^{-4} relationship, is responsible for the wavelength dependence shown by the curves. The introduction of a reflecting surface at the base of the atmosphere modifies the skylight, because of the surface-reflected radiation which is scattered back down by the overlying atmosphere. Curves of intensity of skylight for the Rayleigh atmosphere superimposed on a Lambert surface of albedo $A = 0.80$, which corresponds closely to the albedo of a new snow surface, are shown for the same wave-

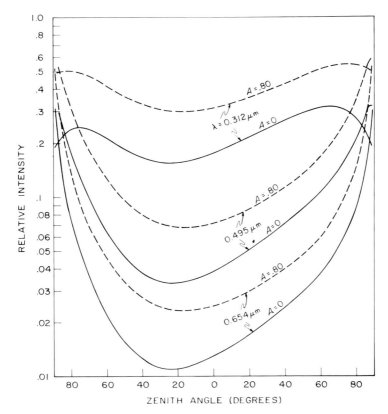

Fig. 4.1 Relative intensity of skylight in the principal plane at a sea-level altitude in a Rayleigh atmosphere for three different wavelengths, two values of surface albedo, and a solar zenith angle of 53.1° (data from Coulson, Dave and Sekera, 1960).

lengths by the dashed curves of Fig. 4.1. A very significant relative increase due to surface reflection occurs at all wavelengths but, of course, the increase is of the largest magnitude at the short end of the spectrum.

Scattering by the Rayleigh atmosphere and reflection from the underlying surface also direct solar radiation back to space from the top of the atmosphere, and thereby contributes to the Earth's planetary albedo. A typical example of the intensity distribution of skylight and the light emerging from the top of the model atmosphere of optical thickness $\tau = 0.25$, corresponding to a wavelength of 0.440 μm, is given in Fig. 4.2 for a solar zenith angle of 53.1° and three different values of surface reflectivity. The similarity of the curves representing the upward and downward streams of light for $A = 0$ is attributable directly to scattering according to the

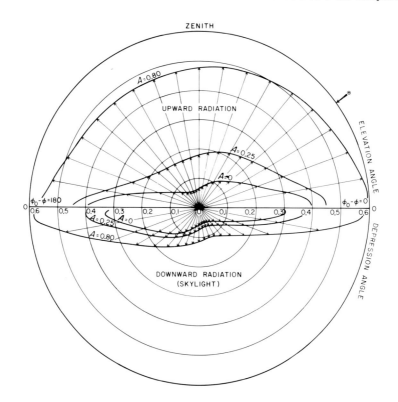

Fig. 4.2 Relative intensity in the principal plane of skylight and of light directed upward to space from the top of a Rayleigh atmosphere at a wavelength of 0.440 μm, solar zenith angle of 53.1°, and three different values of surface albedo ($\tau = 0.25$, $\mu_0 = 0.60$, $A = 0, 0.25, 0.80$). The Sun's position is indicated by the arrow (data from Coulson, Dave, and Sekera, 1960).

Rayleigh phase function, the minimum in each case occurring at about 90° from the direction of the Sun. For $\tau \to 0$, the two streams must be identical but, for this case of moderately large optical thickness, multiple scattering and attenuation effects cause the upward intensity to be the greater. Surface reflectance is seen to have a much stronger influence on the intensity of the upward radiation than on the intensity of skylight.

By the method introduced by Deirmendjian and Sekera (1954), it is a simple matter to compute the total flux of radiation due to skylight incident on a horizontal surface at the bottom of the atmosphere. Typical results are shown by Fig. 4.3. The dashed curve is the flux of skylight due to atmospheric scattering only. The wavelengths corresponding to the optical thicknesses at which computations were made are given in the following tabulation.

τ	λ (μm)	τ	λ (μm)
0.01	0.974	0.15	0.498
0.02	0.821	0.25	0.440
0.05	0.654	0.50	0.374
0.10	0.551	1.00	0.316

For a highly reflecting surface, such as new snow, the contribution due to the backscatter of surface-reflected sunlight may well exceed that due to atmsopheric scattering in the absence of surface reflections for this case of a zenith Sun. The surface effect is less pronounced at other solar

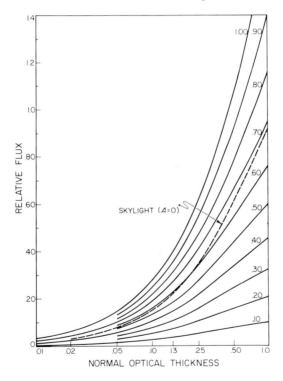

Fig. 4.3 Hemispheric flux of solar radiation which is scattered into the downward direction after being reflected from the underlying surface, as a function of normal optical thickness for various values of surface albedo. The dashed curve is the flux due to atmospheric scattering of the nonreflected incident radiation ($\Theta_0 = 0°$).

angles. At a solar zenith angle of 53.1°, the surface albedo must approach 1.0 before the intensity of backscattered surface-reflected light is equal to the diffusely transmitted light, and for a zenith angle of 78.5° there is no possibility of equivalence of the two.

4.1.2   Case of a Turbid Atmosphere

Scattering by haze, smog, and other aerosol particles has a marked effect on the skylight intensity, the effect being most pronounced in the region of the solar aureole (bright area around the Sun). Kano (1964) has computed the intensity distribution for two simplified models of the turbid

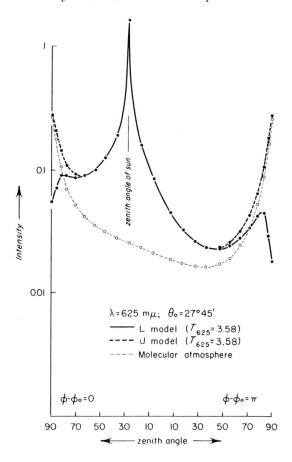

Fig. 4.4   Relative intensity of skylight in the principal plane for a molecular (Rayleigh) atmosphere and for a turbid layer below (L model) and above (U model) the molecular atmosphere. T is the Linke turbidity factor (after Kano, 1964).

atmosphere and obtained the results shown in Fig. 4.4 for $\lambda = 0.625$ μm. The turbid layer was assumed to be below the Rayleigh atmosphere in the L model, and above it in the U model. The aerosol, the Linke turbidity factor of which was 3.58, was assumed to exhibit single scattering properties according to the Mie theory, while multiple scattering was taken into account for the Rayleigh atmosphere. The figure shows the effects of very strong forward scattering by the aerosol particles; the intensity in the vicinity of the Sun is nearly two orders of magnitude higher for the turbid atmosphere than for the molecular atmosphere. Low-level turbidity causes pronounced horizon darkening, which is not present for the upper turbid layer or the molecular atmosphere.

Measurements of the skylight intensity made by Coulson (1969) in the atmosphere of Los Angeles tend to confirm the general features obtained by Kano, but some important discrepancies occur. The intensity distribution for $\lambda = 0.365$ μm for a relatively clear case and for a case of heavy smog, both for downtown Los Angeles in September 1968, are shown in Fig. 4.5. The solar aureole, which is sharply defined for the relatively clear case, is spread out to encompass most of the sky under polluted con-

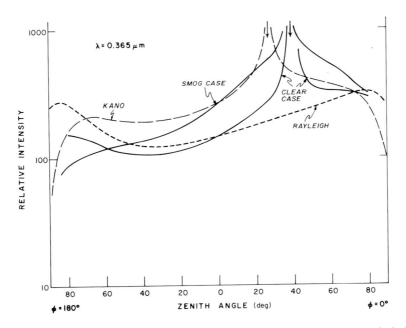

Fig. 4.5 Relative intensity of skylight as measured in Los Angeles for a relatively clear atmosphere, for a polluted atmosphere, and as computed for a Rayleigh atmosphere. The intensity for the Rayleigh case is normalized to that of the clear case at the zenith (data from Coulson, 1969).

ditions, and the intensity is correspondingly increased. Only near the horizons does pollution cause a decrease of skylight intensity. Similar data taken during scans of the sky at different azimuths show that, in polluted conditions, the skylight intensity increases monotonically from the horizon to the Sun at all azimuths, and that the most important single parameter in determining the intensity distribution is the angle at which the light is scattered from the direction of the Sun.

The total global radiation of a given wavelength, which is incident on a horizontal surface, is given by the relation

$$F(\lambda) = F_0(\lambda) e^{-\tau \sec \theta_0} \cos \theta_0 + \int_0^{2\pi} \int_0^{\pi/2} I(\lambda) \cos \theta \sin \theta \, d\theta \, d\phi \qquad (4.5)$$

where the two terms on the right-hand side of the equation are, respectively, the flux on a horizontal surface due to the attenuated direct solar beam and the flux of scattered radiation obtained by an integration of intensity over the hemisphere of the sky.

4.1.3   Case of a Cloudy Atmosphere

There are, in general, two different methods which have been used by various authors in studying the effect of clouds on solar radiation incident on horizontal or inclined surfaces at the ground. The first is an empirical method based on the statistics of various cloud parameters, such as cloud type, thickness, extent of sky covered, and latitude or mean altitude of the Sun. The information derived by this empirical-statistical method is a long-term average of radiative flux for typical cloud conditions at a given location, or a general relation which can be applied under known cloud conditions at any location, to yield a rough value of the resulting flux of radiation. The second approach is that of case studies, either experimental or theoretical, to determine the effects on the radiation regime of specific clouds or cloud types with known or assumed physical parameters. Results obtained from each method are briefly indicated below.

Typical of the empirical-statistical method is that suggested by A. K. Ångström (1924), in which the total solar flux Q on a horizontal surface (the global radiation) is given in terms of the flux Q_c for a cloudless sky and the fraction of possible sunshine duration S by the relation

$$Q = Q_c[k + (1 - k)S] \qquad (4.6)$$

where the empirical constant k is the ratio of Q for the case of overcast to Q for a completely clear sky. A. K. Ångström himself noted that the relation is valid only for climatological means and does not apply directly for individual cases. Other applications of the empirical method have been

TABLE 4.1

Ratio of the Flux of Solar Radiation on a Horizontal Surface for the Case of Overcast with Clouds of various Types to That for the Completely Clear Atmosphere, for Different Values of Air Mass, Expressed as Percent[a]

Air mass	Type of cloud							
	Cirrus	Cirro-stratus	Alto-cumulus	Alto-stratus	Strato-cumulus	Stratus	Nimbo-stratus	Fog
1.1	85	84	52	41	35	25	15	17
1.5	84	81	51	41	34	25	17	17
2.0	84	78	50	41	34	25	19	17
2.5	83	74	49	41	33	25	21	18
3.0	82	71	47	41	32	24	25	18
3.5	81	68	46	41	31	24	—	18
4.0	80	65	45	41	31	—	—	18
4.5	—	—	—	—	30	—	—	19
5.0	—	—	—	—	29	—	—	19

[a] After Haurwitz, 1948.

made by Kimball and Hand (1922), Moon and Spencer (1942), Kasten *et al.* (1959), Budyko (1963), Pochop *et al.* (1968), Quinn and Burt (1968), Greschenko (1968), and others. From 8 years of radiation measurements at Blue Hill Observatory, Haurwitz (1948) determined the ratio of the global radiation Q, for the sky completely covered by clouds of different types, to the global radiation Q_c for the clear atmosphere as a function of air mass m. The results obtained by Haurwitz are summarized in Table 4.1.

The second approach to studies of the effects of clouds, that of computations of the radiative effects for specific models and of measurements for specific clouds, has been pursued by many authors, including Fritz (1955), Feigelson *et al.* (1960), Neiburger (1949), Deirmendjian (1964), Feigelson (1964), Samuelson (1965), Twomey *et al.* (1967), and others. The Monte Carlo method, in combination with the Mie theory of scattering, has recently been applied with considerable success by Plass and Kattawar (1968), for computing the intensity of light transmitted and reflected by clouds. Typical results of their computations for a cumulus cloud model of optical thickness $\tau = 10$ are shown in Fig. 4.6, along with curves of the distribution

$$I \sim \tfrac{1}{3}(1 + 2\cos\theta) \tag{4.7}$$

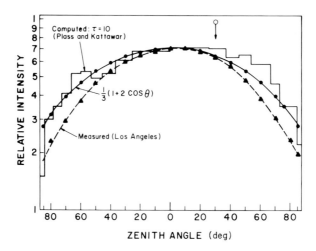

Fig. 4.6 Relative intensity of light transmitted by clouds as a function of zenith angle in the principal plane according to computations and as measured under a uniform stratus overcast (computations: $\lambda = 0.70 \ \mu$m; measurements: $\lambda = 0.65 \ \mu$m, $\theta_0 = 30°$).

suggested by Moon and Spencer (1942) for overcast skies and that measured by Coulson (1971) under a uniform stratus overcast in Los Angeles in September 1969. The curves were all normalized to the same value at the zenith. The general distribution obtained by the Monte Carlo method corresponds as closely to the measurements as could be expected, in view of the fact that the measurements were made under a stratus cloud of unknown optical thickness and size-frequency distribution of the droplets, whereas the calculations were for a model of a cumulus cloud. Furthermore, the Monte Carlo technique is subject to considerable statistical error unless an extremely large number of scattering events are computed, a process which requires large amounts of computer time. The distribution given by $I \sim \frac{1}{3}(1 + 2 \cos \theta)$ corresponds closely to the computed distribution of Plass and Kattawar, but they both give less horizon darkening than was observed. The observations are represented very well, however, by the relation $I \sim \frac{1}{4}(1 + 2 \cos \theta)$. This is a fortunate circumstance, as a simple distribution of skylight is convenient for use in the design of windows for the illumination of working areas in buildings, under unfavorable lighting conditions (Moon and Spencer, 1942).

4.2 MEASUREMENT OF DIFFUSE AND GLOBAL RADIATION

For purposes such as energy-balance studies, the response of organisms to light, directional effects in the atmosphere, and many others, it is desirable to measure both the diffuse solar energy from the sunlit sky and the total flux of solar energy incident on a horizontal surface (i.e., the global radiation). Because of the diffuse character of the skylight, an integration over the entire hemisphere of the sky is required for both diffuse and global radiation measurements. This angular integration imposes stringent requirements on both materials and basic design of pyranometers.

One of the most common errors of pyranometry, and perhaps the most difficult to eliminate, is the "cosine effect" produced by the instrument response being a function of angle of incidence of the radiation. It is an observed fact that the reflectance and absorptance of surfaces in general are dependent on the angle at which radiation strikes the surface. This angular dependence is particularly pronounced for water surfaces. The glint of the Sun reflected from the water is much brighter near sunset than near noon. A similar, though much less pronounced, dependence of the reflectance of paints or other material used to coat the sensor elements of pyranometers may be responsible for significant errors in radiation measurements made with such instruments.

The total global flux G of solar energy which is incident on a horizontal surface, such as the sensing element of a pyranometer, consists of two components—the direct, essentially monodirectional, solar radiation flux F_D incident from a specific solar zenith angle θ_0, and the flux F_d of diffuse skylight which is incident from every possible zenith angle θ and azimuth angle ϕ over the hemisphere of the sky. Thus

$$G = F_D + F_d \qquad (4.8)$$

If T is the effective optical thickness of the atmospheric path from the top of the atmosphere to the level of the sensor, we can write

$$F_D = F_0 e^{-T} \cos \theta_0 \qquad (4.9)$$

where F_0 is the flux across a unit surface, oriented normal to the solar beam, at the top of the atmosphere. The analogous expression for F_d is

$$F_d = \int_0^{2\pi} \int_0^{\pi/2} I(\theta, \phi) \sin \theta \cos \theta \, d\theta \, d\phi \qquad (4.10)$$

where $I(\theta, \phi)$ is the skylight intensity from the (θ, ϕ) direction. Obviously, throughout this section, we are assuming an integration over the entire solar spectrum.

For the ideal case, in which the absorptance α of the radiation sensor is independent of the angle of incidence of the radiation, the energy flux absorbed by the sensor is simply

$$\alpha G = \alpha F_D + \alpha F_d \qquad (4.11)$$

Unfortunately, for any real paints, soot, metal oxides, or other materials which have been found for covering the radiation sensors, the absorptance is not independent of the angle of incidence: $\alpha = \alpha(\theta \text{ or } \theta_0)$. This so-called "cosine effect" can be a significant source of error in measurements made with pyranometers. It is particularly serious for the larger angles of incidence ($\theta > 80°$), where errors of 25% or more are not uncommon among the older types of pyranometers.

The dependence of sensor absorptance on angle of incidence is not the only cause of deviations of instrument response from the true cosine relationship. Striations or other defects in the glass hemispherical envelope(s) of the pyranometer, incorrect leveling, curvature of the receiver surface, or internal reflections inside the instrument may also contribute to error in the radiation measurements.

Various methods have been used to either minimize or compensate for the cosine error of pyranometers. By calibrating instrument response versus angle of incidence of the radiation, it is possible to introduce a correction to the data for the angle of the direct solar flux F_D. However, the distribution of the intensity of skylight varies greatly with the position of the Sun, atmospheric turbidity, and cloudiness, and so the introduction of a cosine correction for the diffuse flux F_d is not ordinarily feasible as a standard operational procedure. Fortunately, the diffuse character of F_d, in combination with the fact that the radiation density of such light incident on a horizontal surface is proportional to the cosine of its zenith angle, assures that the errors arising from the lack of a true cosine response of the receiver are generally of a minor nature for F_d. Furthermore the relative contribution of F_d is a maximum at low solar elevations, when the cosine error is greatest for F_D. Thus, the residual errors remaining after the application of a cosine correction for F_D are probably of minor significance when considered with other errors of the measurements.

However, the computation of cosine errors and the application of corrections so derived is troublesome. It is much more satisfactory to construct the pyranometer to be, automatically, as self-compensating as possible in this respect. Two methods have been introduced for providing such compensation. The earlier one developed for Foster (1951), of the former U.S. Weather Bureau (now National Weather Service), entails the use of a specially shaped diffusing disk, located over the sensor, so that the flux density received by the surface at large angles of incidence is increased over that received by a perfectly flat sensor surface. This method has been successfully employed in the Eppley ultraviolet pyranometer (Drummond and Wade, 1969), the Eppley illuminometer, and in the solar cell pyranometer of Kerr *et al.* (1967). The second method is more directly applicable to the measurement of global and diffuse radiation. Essentially, this design entails arrangement of the radiation sensing surface-hemispherical envelope system in such manner that there is near compensation between energy lost by specular reflection and that gained through refraction effects at the glass surfaces.

4.2.1 Early Development of Pyranometers

Callendar Pyranometers The Callendar pyranometer was invented by the English physicist, H. S. Callendar, in 1898, and an improved version was brought out in 1905. The sensor was made up of four platinum wire grids wound on strips of mica and connected in pairs such that each pair con-

Fig. 4.7 Photograph of the Callendar pyranometer (copyright by The Science Museum, London; reproduced by permission).

stituted an electrical resistance thermometer. One pair of grids was coated with black enamel to promote the absorption of sunlight, while the other pair was left as highly reflecting platinum metal. The grids were arranged in a checker-board configuration and mounted in the evacuated glass bulb which can be seen in the photograph of Fig. 4.7. The total area covered by the grids was a square of dimensions 5.8 × 5.8 cm and the glass bulb was about 9.1 cm external diameter. The difference in temperature developed between the pairs of grids on exposure to solar radiation was, according to Callendar and Fowler (1906), very nearly proportional to the intensity of the radiation incident on them. The temperature difference was recorded by connecting the two pairs of grids as two arms of a self-adjusting Wheatstone bridge with a pen arrangement which made a trace on a revolving drum. Each instrument was accompanied by a calibration factor by which the recorded temperature difference could be interpreted in terms of the flux of radiant energy to which the instrument was exposed. The calibration factor which was furnished was assumed to be independent of intensity and direction from which the radiation came, wavelength of the radiation, and temperature of the grids.

The Callendar pyranometer enjoyed a considerable popularity during the first two decades of this century. Its relatively widespread use was due to its capability of measuring both direct and diffuse radiation, and to its self-recording feature. The first Callendar recording pyranometer used by the U.S. Weather Bureau was secured in 1908 (Covert, 1925), and a number of others were added a few years later. With the development of thermoelectric sensors, the Callendar instruments were gradually replaced. The Weather Bureau had four in operation in 1925, and of the fifteen network stations in operation in 1941, only the Miami station was still using the Callendar electrical resistance pyranometer (Hand, 1941).

Ångström Pyranometer   The electrical compensation principle used by Knut Ångström in the Ångström pyrheliometer during the final years of the nineteenth century was the basis of the pyranometer developed by A. K. Ångström about two decades later (A. K. Ångström, 1919). The design of the pyranometer can be seen from Fig. 4.8. The receiver surface consisted of two white strips (aa) which were coated with magnesium oxide and two black strips (bb) which were covered with platinum black. The four strips were mounted side by side on an insulating framework in the end of a nickel-plated cylinder, and thermojunctions were attached in good thermal contact to the backs of the strips, but insulated from them electrically. The white strips were connected through a sensitive milliammeter and variable resistance to a low-voltage battery. The glass hemisphere (g) was supported by the metal disk (d), into which was inserted a vial of drying agent.

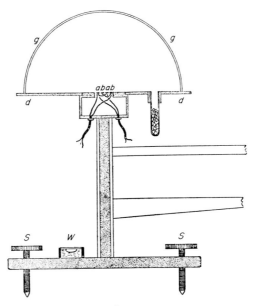

Fig. 4.8   Schematic diagram of the Ångström electrical compensation pyranometer. a, white strips; b, black strips; g, glass hemisphere mounted in disk d; S, adjusting screws; W, spirit level.

The operation of the pyranometer was the same as for any of the compensation-type instruments. The strips were maintained at the same temperature by heat being supplied electrically to the white strips to compensate for the greater amount of radiant energy being absorbed by the black strips.

In spite of some attractive features, the Ångström pyranometer was never put into widespread use.

Smithsonian Pyranometer   Beginning in 1913, an instrument, which appears to have been the first specifically designed for measuring the total Sun and sky radiation falling on a horizontal surface, was developed by the Smithsonian Institution (Abbot and Aldrich, 1916a). The name "pyranometer" was coined for the instrument by Abbot and his colleagues by combining the Greek words for "fire," "up," and "a measure," the resulting term signifying "that which measures heat above." It was hoped that the device would be a standard instrument for solar radiation measurements, and it would be suitable for measurements of nocturnal radiation as well. A total of six models of the pyranometer were constructed, the last two of which were considered satisfactory for radiation measurements.

Details of the final design of the pyranometer were given by Abbot and Aldrich (1916a). The sensor consisted of two blackened manganin strips, the exposed surfaces of which were 6 mm long and 2 mm wide. Strip b was ten times as thick as strip a, and therefore had a higher heat conductance than did strip a. The ends of both strips were soldered to copper posts which were in good thermal contact with a nickel-plated copper block. The strips were electrically insulated, however, from the copper block and from each other by thin mica strips. Tellurium–platinum thermoelements were attached by this waxed paper to the backs of the manganin strips and were connected in series through a sensitive galvanometer, so that the amount of deflection of the galvanometer indicated the difference in temperature of the strips.

Because of galvanometer drifts and other problems, the Smithsonian pyranometer was never put into general use.

4.2.2 Important Modern Pyranometers

The World Meteorological Organization (1965) has defined three classes of pyranometers on the basis of their accuracy and overall system performance. All of the pyranometers which have been developed require calibration with respect to a primary radiation standard, so none can be classed as a standard pyranometer. The bases for the classification are given in Table 4.2.

TABLE 4.2

Classification of Pyranometers

	1st class	2nd class	3rd class
Sensitivity (mW cm^{-2})	± 0.1	± 0.5	± 1.0
Stability (% change per year)	± 1	± 2	± 5
Temperature (maximum error due to changes of ambient temperature—%)	± 1	± 2	± 5
Selectivity (maximum error due to departure from assumed spectral response—%)	± 1	± 2	± 5
Linearity (maximum error due to nonlinearity not accounted for—%)	± 1	± 2	± 3
Time constant (maximum)	25 sec	1 min	4 min
Cosine response (deviation from that assumed, taken at Sun elevation 10° on clear day—%)	± 3	± 5–7	± 10
Azimuth response (deviation from that assumed, taken on clear day—%)	± 3	± 5–7	± 10

The rating, according to these criteria, of the pyranometers which were available in 1965 is the following:

First class
    Selected thermopile pyranometers
Second class
    Moll-Gorczynski (Kipp) pyranometer
    Eppley pryanometer (also called 180° pyrheliometer)
    Volacine thermopile pyranometer
    Dirmhirn–Sauberer (star) pyranometer
    Yanishevsky thermoelectric pyranometer
    Spherical Bellani pyranometer
Third class
    Robitzsch bimetallic pyranometer

Since 1965 several new types of pyranometers have been developed, two by the Eppley Laboratories being based on thermopiles and others by various workers utilizing silicon solar cells as sensors.

The precision spectral pyranometer of Eppley meets the criteria listed for a first class pyranometer, while the Eppley black-and-white pyranometer would fall in the second class category. Instruments based on solar cells, because of their restricted spectral sensitivity and general performance, would be third class pyranometers. A specialized type of pyranometer, the ultraviolet pyranometer developed by the Eppley Laboratories, utilizes a selenium photocell in combination with a quartz diffuser and spectral filter for measurements in selected regions of the ultraviolet. The characteristics of the ultraviolet pyranometer are included in Chapter 5, while those listed previously are discussed in the remainder of this chapter.

Eppley Pyranometer (180° Pyrheliometer)   The pyranometer manufactured by the Eppley Laboratories, Inc., has been by far the most widely used solar radiation instrument in the United States. In Europe, the Moll–Gorczynski (Kipp and Zonen) and the Robitzsch instruments are more common. In a recent survey, Lof *et al.* (1965) tabulated the following world distribution of Eppley pyranometers in routine use for solar radiation observations:

North America	128	Asia	7
South America	0	Australia	0
Europe	4	Antarctica	8
Africa	4	Oceans and Islands	13

The survey showed only nine non-Eppley pyranometers in the North

American observational network, five of which were of the Moll–Gorczynski type from Kipp and Zonen. Undoubtedly there are many more Eppley instruments being used in nonroutine applications, so a knowledge of their characteristics is of major importance in the United States.

A picture of the well-known Eppley pyranometer (variously called pyrheliometer or 180° pyrheliometer) is shown in Fig. 4.9. Although the device is no longer in production, there are large numbers still in operation. The original design of the instrument was developed by the joint efforts of the U.S. Weather Bureau and the National Bureau of Standards (Kimball and Hobbs, 1923). The annular ring construction was suggested by C. F. Marvin, Chief of the Weather Bureau until about 1930. The first pyranometer made with the Weather Bureau thermopile by the Eppley Laboratories was exhibited at the meeting of the American Meteorological Society in May, 1930, and the Eppley Laboratories manufactured the instruments on a continuing basis from that date until they were supplanted by the precision spectral and black-and-white pyranometers in recent years.

Fig. 4.9 The Eppley pyranometer which has received widespread application, particularly in the United States (courtesy of the Eppley Laboratories). This model is no longer in production.

The sensor of this older instrument is a thermopile of gold–palladium and platinum–rhodium alloys, the junctions of which are in good thermal contact but electrically insulated from the concentric rings of the disk shown in the photograph. The disk, with a total diameter of about 29 mm, is made up of two thin, flat, silver rings and a small inside disk, all of which are thermally insulated from each other. The outside ring and the small disk are coated with magnesium oxide, a material which has a high reflectance for radiation in the solar wavelength range. The inside ring is coated with Parsons Optical Black lacquer, and therefore has a low reflectance at solar wavelengths. Because of their different reflectances, the black and white rings assume different temperatures on exposure to solar radiation, and thereby cause an electromotive force to be produced by the attached thermojunctions. The temperature difference between the hot and cold junctions is a function of the total amount of short-wave radiation on the device, so by proper calibration the voltage from the thermopile can be interpreted in terms of the incident solar energy. To a close approximation, the output voltage is a linear function of the incident energy.

Two types of the older pyranometer were produced. They are usually designated as the 10-junction and 50-junction models, although 16 junctions were used in the 10-junction model. Specifications representative of the two models, as given by the manufacturer, are shown in Table 4.3.

The glass bulb surrounding the sensing disk was filled with dry air at atmospheric pressure at the time of manufacture. The bulb itself is approximately spherical, 76 mm in diameter, and made of soda lime glass of approximately 0.6 mm thickness. As can be seen from a typical* curve of transmission versus wavelengths for the glass, shown superimposed on the spectral curve for solar radiation in Fig. 4.10, the effects due to selective transmission of the glass for solar wavelengths are minor. Transmission

TABLE 4.3

Manufacturer's Specifications for The Two Older Models of The Eppley Pyranometer

Characteristic	10-Junction model	50-Junction model
Sensitivity (mV cal^{-1} cm^{-2} min^{-1})	2.5	7.5
Resistance (Ω)	40	100
Time for 98% response (sec)	20	30

* The manufacturer notes that: "This glass is not controlled for transmission, and individual bulbs may vary," (Bull. No. 2, undated).

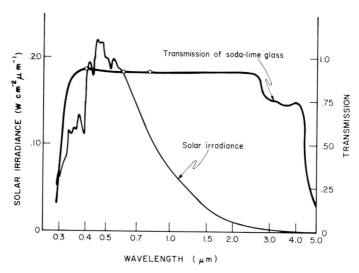

Fig. 4.10 Transmission characteristics of the soda–lime glass envelope of the older model Eppley pyranometer compared to the spectral distribution of solar radiation.

of the glass is almost constant at slightly over 0.90 between about 0.35 and 2.6 μm. According to Thekaekara (1970), 92.15% of the extraterrestrial solar irradiance falls within that wavelength interval and 97.9% of the solar irradiance is between the 50% transmission points at 0.305 and 4.4 μm. Thus the instrument will respond somewhat selectively with wavelength near the effective limits of the solar spectrum, but to the extent that the measurement conditions are similar to the conditions under which the instrument is calibrated, the spectral selectivity will introduce no errors into the measurement results. For practical purposes the effects of spectral selectivity due to the glass are probably negligible.

The (uncompensated) Eppley pyranometers are subject to a significant dependence of sensitivity on the temperature of the instrument. Unfortunately, that factor has been largely neglected in the published data from the radiation measurements network. The sensitivity decreases with increasing temperature by between 0.05 and 0.15% per 1°C rise in temperature over the range −50 to +40°C. Since the normal response of thermocouples is for an increase of sensitivity with increasing temperature, it appears that the observed decrease is caused by convection inside the pyrheliometer bulb (MacDonald, 1951). The results of three different tests of the change of instrument response with temperature for a constant value of irradiance is shown by the curves of Fig. 4.11. The data are averages for two instruments tested by the National Bureau of Standards, five by the Weather Bureau, and two by Drummond (1965).

A considerable improvement in the performance of the Eppley instrument can be achieved by the insertion of a thermistor of about 2000 Ω resistance in series with one of the thermopile leads. Since the positive temperature coefficient of the thermistor is much larger in absolute magnitude than is the negative temperature coefficient of the thermopile, an auxiliary shunt with about 1500 Ω resistance is introduced into the circuit. The temperature compensation is brought about by the decrease of resistance, and a consequent decrease of potential drop in the circuit, with increasing temperature of the thermistor. The thermistor is inserted in the heat sink of the receiving element and thereby subjected to the temperature of the element while being shielded from incident radiation. The improvement in instrument performance which can be obtained by the device is indicated by the fourth curve of Fig. 4.11. The data (Drummond, 1965) are the averages of the results obtained from ten temperature-compensated

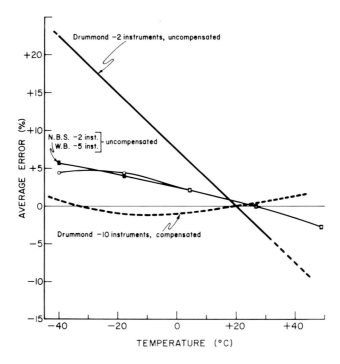

Fig. 4.11   Average error as a function of temperature for Eppley pyranometers which are compensated and uncompensated for temperature effects.

instruments on exposure to incident radiation of 1.2 cal cm^{-2} min^{-1}. They
have been normalized to indicate zero error at 20°C. Similarly small errors
due to temperature effects can be achieved over a wider temperature range
(e.g., -80 to $+50$°C) by a circuit utilizing components made of silicon,
a material which has a greater positive temperature coefficient than that
of manganin.

The Eppley instrument, as well as others, is subject to an additional
error associated with changing angle of incidence of the radiation on the
detector. Results of tests reported by MacDonald (1951) and listed in
Table 4.4 show that errors from this cause vary considerably among different instruments at all angles, but the instrument response is degraded very
seriously at angles of incidence greater than about 60°. The reason for the
degradation is probably a change of reflectance of the receiving surfaces,
particularly the black surface, with angle of incidence, although improper
alignment of the elements or inhomogeneities in the glass envelope may be
contributing factors.

For special purposes the Eppley pyranometer may be used in an orientation other than the normal upright position with the plane of the receiver
horizontal and exposed to the upward hemisphere. Data which have been
given by MacDonald (1951) and corroborated by the manufacturer (Bull.
No. 2, 1964) indicate that the instrument operates satisfactorily in an
inverted position, but that the response is decreased by an amount varying
from 2 to 5% when the receiving surface is oriented vertically. Convection

TABLE 4.4

*Errors Caused by Instrument Response Not Being Directly Proportional to the Cosine
of the Angle of Incidence of the Radiation on the Detector[a]*

Angle of incidence	Pyrheliometer number			
	1754 (%)	1973 (%)	1220 (%)	1221 (%)
0	0	0	0	0
30	+3	0	+2	+2
60	+5	+1	0	−1
70	—	—	−6	−4
75	+1	—	—	—
80	—	−21	−19	−18
85	+3	—	—	—

[a] After MacDonald, 1951.

currents set up inside the glass envelope are probably the cause of the decreased response.

Drummond (1965) estimates that accuracies of the order of ±2 to 3% are attainable for daily summations of radiation with the temperature compensated Eppley pyranometers. Individual hourly summations, even with the most carefully calibrated equipment, may be in error by 5% or more, particularly at the lower Sun elevations. It is not unreasonable to expect these errors to be at least doubled for routine observations with equipment which is reasonably well maintained. Isolated cases of poorly maintained equipment, but still in the regular observational network, have shown systematic errors of more than 10% for monthly averages, with the shorter period errors being superimposed on the large average value. It is perhaps a platitude to recommend frequent calibration and competent maintenance of the instrumentation, but these factors are probably more important for radiation instruments than for any other instruments used in normal meteorological measurements.

Eppley Black-and-White Pyranometer   The older 10- and 50-junction pyranometers discussed above have been replaced by the "black-and-white" pyranometer shown in Fig. 4.12. The detector in this new instrument is a wire-wound thermopile made by electroplating copper on con-

Fig. 4.12   The Eppley "black-and-white" pyranometer (courtesy of the Eppley Laboratories).

stantan. The hot and cold junctions are painted with Parsons Optical
Black paint and barium sulfate, respectively. According to specifications
by the manufacturer, the built-in temperature compensation provides a
signal which is independent of temperature to within $\pm 1.5\%$ from -20 to
$+40°C$. The sensitivity is about 7.5 mV/cal cm^{-2} min^{-1}, and the deviation
from a true cosine response is within $\pm 2\%$ for incident angles of 0° to
80°. An important improvement over the older device is the provision of
an optically ground Schott WG 7 glass envelope, in place of the blown
glass bulb used previously. The glass is removable for ease in cleaning and
repair of the sensing surface.

Eppley Precision Spectral Pyranometer   The first model of this instrument
was introduced in 1957 (Marchgraber and Drummond, 1960). The principal improvements over the earlier Eppley instrument were (a) electrical
compensation for the dependence of sensitivity on ambient temperature,
(b) optical compensation for deviation of response from the cosine law,
and (c) provision for the use of broadband spectral filters. The detector of
the original model was a conventional type of bismuth–silver thermopile
of 15 junctions yielding about 7 mV/cal cm^{-2} min^{-1}. A second version of
the precision instrument, which is the model in current production, is
considerably smaller in size than the original version and utilizes a thermopile of copper electroplated on constantan wire over one-half of each turn
of a wire-wound thermopile. The design of the current model is shown in
the cross-sectional diagram of Fig. 4.13.

Specifications supplied by the manufacturer (1973) are given in the

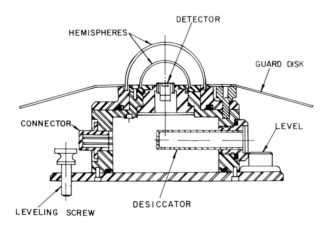

Fig. 4.13 Cross-sectional diagram of the current model of the Eppley precision
spectral pyranometer (courtesy of the Eppley Laboratories).

following tabulation:

Sensitivity	5 mV/cal cm^{-2} min^{-1} (approx.)
Impedance	300 Ω
Temperature dependence	$\pm$ 0.5% from -20 to $\pm$ 40°C
Cosine error	$\pm$ 1% over angles of 0° to 80°
Response time $(1/e)$	1 sec
Linear response up to	4 cal cm^{-2} min^{-1}

One attractive feature of the device is the possibility of replacing the outer one of the two glass hemispheres with Schott color filters for measurements of solar radiation in selected spectral bands. The filters available in the required hemispherical configuration have lower wavelength cutoffs at 0.5 μm (filter GG 14), 0.53 μm (OG 1), 0.63 μm (RG 2), and 0.7 μm (RG 8). Quartz hemispheres are also available for measurements in the ultraviolet.

A feature of the instrument which should be mentioned is its apparent change of sensitivity with the use of the spectral filters (Drummond and Roche, 1965). It was found that an increase of temperature of the filter glass by the absorption of solar radiation, with a consequent heating of the inner-glass hemisphere and modified long-wave radiative distribution at the detector, is responsible for the apparent increase of sensitivity. Although it seems likely that the effect must be a function of meteorological conditions, particularly wind speed, Drummond and Roche (1965) advocate applying the following multiplication factors to the measured values, for either pyranometer or pyrheliometer, whenever the spectral filters are used: OG1: 0.94; RG2: 0.925; RG8: 0.91.

Spectrolab Pyranometer   The model SR-75 Spectrosun pyranometer* is probably the newest entry in commercially available pyranometers. The National Weather Service is introducing a relatively large number of these instruments (over 40 by latest information) into the United States radiation network to replace some of the older pyranometers and upgrade the system.

The model SR-75 is the same in outward appearance as the Eppley precision spectral pyranometer described above. It has two glass domes, the outside one of which can be replaced by absorption optical filters for broad band spectral measurements. The thermopile is covered with 3M Black Velvet paint and has a total resistance of less than 500 Ω. Other

---

* Manufactured by Spectrolab, 12484 Gladstone Ave., Sylmar, Calif. 91342.

specifications, as supplied by the manufacturer, are given in the following tabulation:

Sensitivity:	5 mW/cal cm^{-2} min^{-1}
Temperature error:	± 1% from −20 to + 40°C
	± 2% from −40 to −20°C
Cosine error:	± 3% from 0° to 70° angle of incidence
	± 7% from 70° to 80° angle of incidence
Time of response (1/e):	1.6 sec

Insufficient experience has been gained with this instrument as yet to indicate the extent to which the specifications are met in actual practice or the long-term stability of the device.

Moll–Gorczynski Pyranometer (Solarimeter)   The Moll–Gorczynski pyranometer (often called the solarimeter) is based on the thermopile designed by Dr. W. J. Moll (1923), Lecturer in physics at the University of Utrecht. A Moll thermopile was used by Dr. Ladislas Gorczynski, Director of the Polish Meteorological Institute, for constructing both a pyrheliometer and a pyranometer in about 1924. The thermopile for the early instruments consisted of no less than 80 thermocouples arranged inside a circle of about 2 cm diameter, and gave up to 16 mV for incident radiation of 1 g cal cm^{-2} min^{-1}. The rapid response of the thermopile to changes of incident radiation, the very small dependence on ambient temperature, and the essentially linear relationship between radiation intensity and instrument response all boded well for the future of the Moll thermopile in radiation measurements. As will be seen below, those bright prospects of the early

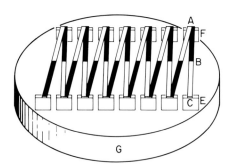

Fig. 4.14   Diagram of the Moll thermopile. B, active junctions of metal strips; A and C, passive junctions of metal strips; E and F, copper mounting posts; G, massive brass plate.

days have not been completely reflected in the number of instruments using the Moll thermopile which are currently in use.

The basic design of the Moll thermopile can be seen from the diagram of Fig. 4.14. The thermojunctions are made of very thin (0.005 mm) blackened strips of manganin and constantan which are joined at point B and soldered at points A and C to copper posts E and F. The mounting posts are clamped to a thick brass plate G. A thin coat of lacquer between the posts and plate provides electrical insulation without sacrificing good thermal contact. The active junctions are normally arranged along a line through the center of the plate, and the passive junctions are on top of the copper posts. The small heat capacity of the strips and efficient heat conduction to the copper posts assures a small lag coefficient, and the large heat capacity of the brass plate with copper posts assures approximately constant temperature and makes it unnecessary to shield the passive junctions from the incident radiation.

The Moll thermopile is the basis of the Moll–Gorczynski pyranometer (solarimeter), currently manufactured by Kipp and Zonen. A photograph of the instrument in its normally mounted configuration is shown in Fig. 4.15. The Kipp and Zonen instrument enjoys a considerable popularity in

Fig. 4.15 The Moll–Gorczynski pyranometer as manufactured by Kipp and Zonen (courtesy of Kipp and Zonen).

Europe, and somewhat less in Africa and Asia. The survey of radiation instrumentation by Lof *et al.* (1965) shows 65 of the 219 pyranometers in the radiation network of Europe to be Kipp and Zonen instruments, although even there, the simpler and less expensive Robitzsch pyranometers were more numerous (107). Kipp and Zonen pyranometers accounted for 30 out of the total of 71 on the African continent, and 17 of 142 in Asia. They are little used in other parts of the world. For instance, the North American network contained only 5 Kipp and Zonen instruments out of a total of 137. The survey showed that in 1965 there were 2 in use in South America, 1 in Australia, and 2 in Antarctica.

The thermopile of the current model of the Moll–Gorczynski pyranometer consists of 14 blackened manganin-constantan thermoelements arranged in a 12×11-mm rectangular configuration. It is protected from the environment by being covered by two concentric hemispherical glass domes (see Fig. 4.15), which can be removed for inspection and cleaning. Condensation of moisture on the inside of the domes is prevented by the enclosed spaces being connected, through a tube, to a bottle of desiccant, such as silica gel. The 30-cm-diameter radiation shield surrounding the outer dome, and coplanar with the sensitive element, prevents direct solar radiation from heating the base of the instrument. The domes have uniformly high-transmission characteristics throughout the spectral range of $\lambda = 0.32$ μm to about $\lambda = 2.5$ μm (Kipp and Zonen, 1967). The transmission falls off steeply beyond 2.5 μm. Transmission is relatively constant in the ultraviolet at wavelengths $\lambda > 0.32$ μm. There is some decrease below $\lambda = 0.32$ μm, the transmission at $\lambda = 0.30$ μm being down to about 0.68. Approximately 3.0% of the extraterrestrial solar energy is contained in the region of $\lambda > 2.5$ μm, and only about half of the energy normally reaches the ground. Approximately 1.0% of the extraterrestrial energy lies between the ozone cutoff at $\lambda = 0.30$ μm and $\lambda = 0.32$ μm, and strong atmospheric scattering returns about half of that back to space. Since the glass still transmits about 80% in this range, the energy in the ultraviolet which is blocked by the glass is only of the order of 0.1% of the total. Thus, a single glass envelope would cut out not more than 1.6% of the incident solar energy, and a double glass dome about 3.2%. This diminution will, of course, be compensated for in the calibration of the instrument if the radiation used for the calibration has the same spectral distribution as that for which the measurements are made. Thus, for normal use the errors arising from imperfect transmission of the glass are not of major significance.

Sensitivity of the instrument is 8–9 mV/cal cm^{-2} min^{-1}, and the internal resistance of the thermopile is approximately 10 Ω. The temperature coefficient amounts to 0.0015–0.0020 (°C)$^{-1}$ in the sense of decreasing

sensitivity with increasing temperature. No temperature compensation is
provided in the instrument.

Deviations from the exact cosine response with angle of incidence of
the radiation, and changes of response with azimuth of the incident radiation, can be a source of considerable error in the Moll–Gorczynski pyranometer. It appears that the lack of a perfect cosine response for this, as
well as for many other radiation instruments, is mainly brought about by
imperfect reflection or absorption characteristics of the coatings on the
sensitive elements. Specifications by the manufacturer (private communication, 1967) show that deviations from the cosine law will be within
±1% for incidence angles of less than 75°, with greater and individually
variable deviations at larger angles of incidence. Heinzpeter (1952) found
a relatively strong dependence of instrument response on the azimuth of
the source with respect to the sensing surface, the dependence being particularly pronounced at large angles of incidence. His results are represented
by the curves of Fig. 4.16, in which percent error is given as a function of
azimuth for five different angles of incidence. The data have been normalized to indicate no error at zero azimuth and zero angle of incidence. It
appears that the azimuth dependence is produced by the nonsymmetric
orientation of the thermojunctions of the Moll thermopile, a fact which
accounts for the double maximum configuration of the curves.

Dirmhirn and Sauberer (Star) Pyranometer The star pyranometer, also
known as the Dirmhirn–Sauberer pyranometer and the Stern pyranometer,

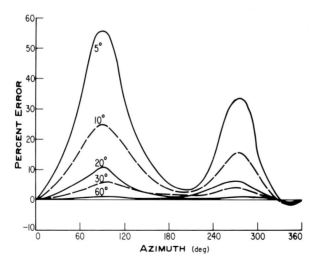

Fig. 4.16 Error of measurement of the Moll–Gorczynski pyranometer as a function
of azimuth at various angles of incidence of the radiation (data from Heinzpeter, 1952).

utilizes 32 (16 in some models) copper plates of 50 μm thickness, half of which are blackened for high absorptivity of radiation and half of which are covered with a highly reflecting white paint. The two sets of plates are mounted as alternate black and white segments radiating as a "star" from a central point, together forming a flat circular disk of approximately 5 cm diameter. Figure 4.17 is a photograph of the star pyranometer (lower left) together with a directly indicating meter and a Dirmhirn radiation balance probe.

The two types of plates are thermally isolated from each other by being mounted on poorly conducting concentric rings which are themselves thermally isolated from the main baseplate of the instrument. Thermojunctions of either copper-constantan or manganin-constantan are soldered to the underneath side of the copper plates, hot junctions being attached to the black segments and cold junctions to the white segments. The resulting 32-junction thermopile generates an emf of 1.8 mV/cal cm^2 min^{-1} (Dirmhirn, 1958) and has an internal resistance of approximately 5 Ω. Response time is such that 98% response to a sudden change of radiation is accomplished in 20–30 sec.

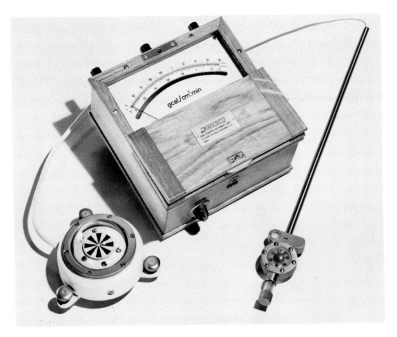

Fig. 4.17 The star pyranometer (lower left) with a directly indicating meter and a radiation balance probe (courtesy of the Kahl Scientific Instrument Corp).

The sensor disk is covered by a ground and polished glass hemisphere of 2 to 3 mm thickness and 110 mm diameter (70 mm in the 16-element models) which transmits more than 90% of the solar radiation through the 0.3- to 3.0-μm spectral range. The sensor section is hermetically sealed, and the enclosed air is in contact with a drying agent which eliminates moisture condensation on the inside of the glass hemisphere. The sensor assembly is enclosed in a circular housing of about 17 cm diameter, with an overall height of 10 cm. The housing is painted white to minimize radiant heating effects.

The response of the star pyranometer has been found to be very stable with time, particularly after an initial 6-month aging period, and by proper calibration against a standard instrument, it yields reliable and consistent radiation measurements. The circular symmetry of the receiving disk eliminates any azimuth dependence of sensitivity, and the measurements available indicate a relatively true cosine response down to an incidence angle of 75° (Dirmhirn, 1958). Instrument response is closely proportional to radiant intensity. Temperature compensation is not included, but the temperature coefficient is sufficiently low as to make temperature compensation unnecessary for most observational conditions. For instance, one model shows only a 1.8% change of sensitivity over a 60°C temperature range.

The star pyranometer has found relatively wide acceptance in various parts of the world. It has been recommended as a suitable instrument for

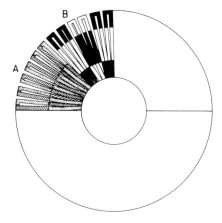

Fig. 4.18 Schematic pattern of black and white segments constituting the thermojunctions of the Yanishevsky pyranometer. A: unpainted section; B: painted section (redrawn from Kondratyev, 1969).

the measurement of global and sky radiation by a commission of the World Meteorological Organization.

Yanishevsky Pyranometer   The Yanishevsky pyranometer is the principal instrument in the U.S.S.R. for measuring diffuse solar radiation, global solar radiation, and surface albedo (Yanishevsky, 1957; Kondratyev, 1965). The sensor is constructed either in a square checkerboard pattern of alternate black and white squares and rectangles or in a radial pattern of alternate black and white segments. The latter of these types is shown in the diagram of Fig. 4.18 and the photo of Fig. 4.19. The thermocouples are composed of alternate strips of manganin and constantan. The hot junctions are blackened with soot and the cold junctions are whitened with magnesium. The sensor assembly is covered with a single glass hemisphere, and an auxiliary opaque hemisphere is provided for obtaining the zero setting of the instrument. Condensation of moisture inside the glass is prevented by the use of a drying agent in a cavity set into the base of the instrument.

The older models of the Yanishevsky pyranometer suffered from a deficiency due to the thickness of the paint on the sensing elements (Yani-

Fig. 4.19 Photograph of the Yanishevsky pyranometer in operation at the Radiation Observatory, University of Moscow (photograph by author).

shevsky, 1972, private communication). The metal strips were very thin but the layer of paint was thick enough to cause a significant roughness to the surface. This roughness, in combination with spaces between the elements, caused the deviation from the cosine law, mentioned above. In the newer models there are no significant spaces between the elements and the paint is of uniform thickness over the whole sensing surface. The construction is of three rings, two of manganin and one of constantan, set close together and cut into narrow radial strips by means of an engine with a fine spark cutter. The radial strips are then painted black and white, as shown, so that an approximately plane surface is obtained.

The Yanishevsky pyranometer is used as a relative instrument, and therefore it requires calibration against a standard. The operational method of calibration is by the sun-shadow method, with the Ångström pyrheliometer as the primary standard. Deviation of the response with solar angle from the ideal cosine law is considerable, and a correction is applied to the measurements for this cosine effect. An additional correction is necessary for wavelength selectivity of the instrument when it is used for measuring only the diffuse radiation, since the spectral distribution of the skylight is considerably different from that of the direct or total global radiation.

The instrument is used in either of two modes of operation for measuring surface albedo. The usual method is to employ a single instrument for the purpose of orienting the receiving surface alternately upward and downward, by means of a special joint in the mount. This procedure is based on the assumption that short-period fluctuations of the incident radiation are of minor significance. Since such fluctuations do sometimes occur, significant errors can be introduced in this mode of operation. The second method is to use two instruments, one facing upward and the other downward, the albedo being simply the ratio of the resulting two measurements. Great care must be exercised in this mode, as a systematic error in one of the instruments results in consistently erroneous albedo determinations. Another disadvantage is that a larger complement of instrumentation is required for this method than for the single instrument method.

Pyranometer of the Physico Meteorological Observatory   This observatory, located in Davos, Switzerland, has suspended the manufacture of pyranometers,* but apparently the instruments are still available. The basic sensor is a 60-element copper-constantan thermopile, which is used in both the pyranometers and pyrradiometers. The main difference between the two types of instruments is in the type of hemispheric shield over the

---

* According to Dr. C. Frolich in June, 1973, a search was then under way to find a manufacturer.

sensor. Two optically polished glass domes are used for the pyranometer, and a single polyethylene dome for the pyrradiometer. Manufacturers specifications of the instruments, together with a photograph of both types, are given in a discussion of the pyrradiometer in Chapter 11.

Robitzsch Bimetallic Pyranometer (Actinograph)   In spite of its low accuracy and generally marginal performance, the Robitzsch-type bimetallic pyranometer (actinograph) has been widely used in various parts of the world. In a survey of instruments employed in regularly reporting solar radiation networks throughout the world, Lof *et al.* (1965) found that about half of the total number of radiation instruments in Europe were of the Robitzsch type. In Asia there were 89 Robitzsch instruments out of a total number of 142. At the time of the survey, South America had 47 Robitzsch instruments, Africa had 16, and Australia had 7 in their regular networks. The survey showed none of the Robitzsch instruments in the regular network of North America, but it is known that there are many instruments of the bimetallic type in use in biological, agricultural, and other environmental investigations in North America.

Fig. 4.20   The Robitzsch bimetallic pyranometer (actinograph) (courtesy of Science Associates).

The popularity of the Robitzsch design is due no doubt to its basic simplicity and to its self-contained, self-recording feature which is particularly desirable for remote operation. However, even with the many improvements which have been incorporated into various models since the first standard design (Robitzsch, 1932), the instrument is not recommended by the Radiation Commission of the International Association of Meteorology for any measurements except daily totals of radiation (CSAGI, 1958, p. 417) and only then "with the reservation that even if a calibration factor varying from month to month is used, these daily totals must be regarded as having an accuracy of not better than ±5–10 percent."

A number of models of the Robitzsch-type instrument are available, but they are all basically similar to that of the photograph shown in Fig. 4.20. The main sensor is a blackened bimetallic strip of dimensions about 8.5×1.5 cm which acts simply as a bimetal thermometer. One end is free to move as changes of temperature cause a distortion of the strip. Through a mechanical linkage, the movement of the free end of the blackened strip causes a deflection of a recorder pen as it is making a trace on a recorder chart mounted on a clock-driven drum. Two standard speeds of rotation of the drum are available, one rotation in 24 hr and one rotation in 7 days. For operation the entire mechanism is enclosed in a sealed, desiccated metal case, the main sensor section being covered by a hemispheric glass dome and a window being provided for viewing the pen and recorder chart.

Since the amount of deflection of the bimetallic strip is determined entirely by its temperature, the instrument is sensitive to the ambient environment as well as to the flux of incident radiation. Many ingenious devices have been employed in the various models to decrease such environmental effects, but none has been completely successful. One method which has been almost universally employed is that of flanking the blackened bimetallic element by two highly reflecting bimetallic elements which move the mount of the blackened element to compensate for its added deflection due to environmental temperature changes. Unfortunately, the compensation by this mechanism is by no means perfect, so there are relatively large residual temperature errors in the record. According to Courvoisier (1954) the dependence of the calibration factor for the SIAP (Società Italíana Apparecchi Precisíone) model can be represented by the relation $K = 19.4 \ (1 \pm 0.003T)$, where temperature T is in degrees Kelvin.

Changes of the calibration constant with changing angle of incidence of the radiation have been found to be very large for the Robitzsch instruments. Because of its distortion with temperature, the blackened sensor strip cannot remain flat, a fact which makes the calibration constant a function of both azimuth and zenith angle of the incident radiation. Additional effects such as imperfections in the glass of the dome and nonuniform concentrations of energy by caustics produced by the glass contribute to the directional sensitivity of the calibration factor. While these effects vary with models, they are known to be very large in some cases, especially in the older designs (Thams, 1943). In one (Feuss) instrument the calibration factor was found to decrease by 40% as the angle of incidence increased from 10° to 70°. Obviously with such a strong dependence of the calibration factor on angle of incidence, it is necessary for even moderate accuracy to apply appropriate diurnal and seasonal corrections to the measurements.

The calibration factor is also a function of the incident energy itself, the factor increasing with an increase of incident radiation. In at least one case (Morikofer and Thams, 1937) the calibration factor changed by as much as 20% over the normal diurnal range of incident solar energy. Although recent modifications have decreased its magnitude, this nonlinear response is not an uncommon characteristic and results in the necessity of additional corrections being applied to the measurements.

The large heat capacity of the bimetallic temperature element and its associated mountings makes a slow response of the instrument to changes of incident energy flux. While the time constant varies among the different individual designs, 10–15 min are normally required for a 98% response to a sudden change of energy. This feature makes the Robitzsch-type instruments unsuitable for measurements involving integration periods shorter than several hours to a day, and it creates great uncertainties in correcting the measurements for temperature effects and for variations due to angle of incidence. Because of the required integrations, an accurate zero setting of the pen is very important in evaluating the records. Either a drift of the zero or an incorrect setting can cause large errors in integrating the area on the chart from which radiation values are determined. Probably the most accurate method of obtaining an accurate zero setting is to cover the glass dome with an opaque cap and allow 15 min or so for the instrument to reach equilibrium before making the zero adjustment. Setting the zero from the nighttime trace does not provide a correct zero position since long-wave radiation exchanges may cause the temperature of the black strip to be different from that of the reference strips.

Spectral transmission of the glass dome is apparently not uniform among the various models available, but it is normally greater than 90% over the range 0.4 to 2.0 μm. In the absence of spectral changes of the incident radiation, this range could give values representative of the entire solar spectral range with proper instrument calibration. The spectral changes which do occur in the atmosphere, particularly outside the range of adequate transmission of the glass dome, are small enough so that errors of measurement due to imperfect transmission by the glass are probably negligible in comparison to those arising from other causes. However, imperfections in the glass itself, particularly in the older models, may cause excessive dependence of the calibration factor on direction from which the radiation is incident on the instrument.

Efforts to increase the accuracy and reliability of the Robitzsch instruments have yielded many modifications of the basic design. Perhaps the most widely used modern designs are those of Casella (London), Feuss (Berlin), and Società Italíana Apparecchi Precisíone (Bologna, Italy), and Foster (Canadian Meteorological Service). In some of the instruments a white or polished reflecting plate is mounted below the main sensing strips, the objective being to decrease heating of the linkage and of the inside of the case. This feature has been eliminated from the Casella design, however, because the radiation incident on, and reflected from, the plate causes excessive variations of response with angle of incidence. Various configurations of radiation shields have been employed around the sensor, and different types of both bimetal elements and sensor coatings have been used. In one model (Casella) auxiliary blackened bimetal strips have been incorporated into the sensor mounting in an effort to improve the temperature compensation and to decrease angular sensitivity.

In summary, we can say that with the best design which has been developed so far, the Robitzsch-type instruments are suitable only for daily totals of radiation in which accuracies of $\pm 10\%$ are adequate. For those undemanding requirements, the Robitzsch design provides a simple and convenient solution. Great caution must be exercised, however, in the choice of a particular design; by no means do all of the Robitzsch instruments on the market yield measurements of $\pm 10\%$ accuracy on a continuing and reliable basis.

Pyranometers Based on Photovoltaic Cells The invention of silicon photovoltaic solar cells at Bell Laboratories in 1954 opened up new possibilities for developing simple and inexpensive radiation measuring instruments. Although the accuracy obtainable with such devices is not high, it is probably adequate for many uses in integrating over periods of a day or longer. Kerr *et al.* (1967) have obtained measurements over a 5-month period in winter in which the solar cell in combination with a mercury current integrator yielded radiation data with a standard error of $\pm 3.8\%$ when compared to that from an Eppley pyranometer. It is likely that measurement errors of the Eppley instrument itself accounted for some of that indicated by the analysis. A 4-month series of measurements with larger radiation fluxes in summer showed a standard error of $\pm 2.8\%$ in comparison to the Eppley data. In both series the errors are for 1-day integration

periods. Kerr *et al.* (1967) estimate a short-period standard error of ±5%
with the solar cell instrument. Other advantages, in addition to simplicity
and low expense, of instruments based on solar cells are their essentially
instantaneous response (about 10 μsec), high current output, direct pro-
portionality between current and incident radiation, and overall stability
with time and exposure to weather.

Inaccuracies in the measurements are brought about by several undesirable characteristics of the silicon solar cells themselves. First, the cells
respond selectively with wavelength of the incident radiation. As is shown
by the plot of relative response versus wavelength for a typical solar cell
(Selcuk and Yellot, 1962) in Fig. 4.21, the response is negligibly small at
wavelengths shorter than 0.40 μm and longer than 1.1 μm and there is a
relatively sharp maximum at about 0.85 μm. This characteristic would be
be of no particular significance if the spectral distribution of the measured
radiation were constant. It could be accounted for by calibrating the instrument in the same situation in which measurements were to be made.
It is known, however, that the spectral distribution of solar radiation which
reaches the ground is not strictly constant, but it varies with solar elevation, cloudiness, turbidity, and water vapor content of the atmosphere,
and reflection characteristics of the underlying surface. This change will
cause an error of perhaps 2% in skylight measurements with solar cells
between the extremes of completely clear to completely overcast skies.

For albedo measurements the spectral response of the instrument
may cause considerably larger errors than for skylight because of the dependence of reflectance on wavelength. The reflectance of mineral soils

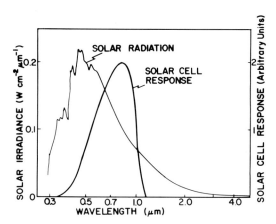

Fig. 4.21 Spectral response of silicon solar cells compared to the spectral distribution
of solar radiation.

usually increases gradually with increasing wavelength, whereas green vegetation shows a dramatic increase of reflectance at wavelengths between 0.7 and 0.8 μm. The incident radiation, however, does not normally have a corresponding increase of intensity at those wavelengths. If A is the spectrally integrated albedo for the interval λ_1 to λ_2, we can write

$$A = \int_{\lambda_1}^{\lambda_2} R_\lambda f_\lambda \phi_\lambda \, d\lambda \Big/ \int_{\lambda_1}^{\lambda_2} f_\lambda \phi_\lambda \, d\lambda \qquad (4.12)$$

where R_λ is the surface reflectance, f_λ is the flux of incident radiation, and ϕ_λ is the instrument response, all of which are wavelength dependent, in general. Only for instruments with a "flat" response, such as that approximated by a blackened thermopile, can ϕ be eliminated from the equation, in which case the wavelength dependence of R_λ and f_λ does not introduce errors into the albedo determination. For silicon cells, germanium cells, and many other types of detectors, however, ϕ_λ is a relatively strong function of λ, and serious errors may be thereby introduced into albedo measurements made with such detectors.

Silicon solar cells are normally used in the short-circuit mode, for which case the current generated by the cell is closely proportional to the incident radiant flux. The high current output (20–30 mA in full noonday Sun) makes it feasible to use current integrators for conveniently integrating over a period of a day or more. Federer and Tanner (1965) combined solar cells with a mercury current integrator to develop the radiation integrating "pyrigrator." The dark current of the measuring cell (which is also integrated) was minimized by connecting a second cell, operating in the voltage mode, as an external bias voltage to the measuring cell.

The temperature coefficient of silicon cells is of the order of 0.0004–0.0010 per °C. It varies somewhat with wavelength of the radiation, being positive in the $0.85 \leq \lambda \leq 1.1$ μm region but near zero or slightly negative for $\lambda < 0.85$ μm. Kerr et al. (1967) consider that an error of no greater than 2% will be introduced into the measurements made at temperatures of from 0 to +40°C if the instrument is calibrated at 20°C, thereby making temperature corrections unnecessary for many applications. However, Selcuk and Yellott (1962) have introduced temperature compensation into their Sol-A-Meter by connecting a low resistance thermistor in parallel with a shunting resistor in the output from the solar cell.

Because of the nature of the surface of silicon solar cells, their response deviates strongly from the ideal cosine law with angle of the incident radiation. Measurements of the cosine response, as obtained by Kerr et al. (1967) and by Selcuk and Yellott (1962) for bare solar cells are shown by Fig. 4.22. Specular-type reflection from the semiglossy surface of the cells

is undoubtedly the cause of the pronounced decrease of response with increase of angle of incidence. By mounting the cell below a contoured plastic
diffuser, Kerr *et al.* (1967) were able to greatly improve the cosine response,
as shown by the upper curve of Fig. 4.22. The price paid for the improvement is a considerably lowered overall sensitivity of the resulting instrument, but the use of the diffuser increases the validity of radiation integrations over whole day periods.

As might be expected, the adaptability and nominal price of silicon
solar cells has encouraged their incorporation into several different configurations of radiation instruments, although it seems that the cost of the
final instrument often does not reflect the moderate cost of the primary
sensor. Various types of covers have been placed over the cells to improve
their cosine response or to make the instrument weatherproof, and solar
cells have been incorporated into both pyranometer and pyrheliometer
types of instruments. One particularly interesting instrument, a photograph of which is shown in Fig. 4.23, uses four solar cells to power an
ampere-hour meter to give an integrated value of the total horizontal
radiant energy in the 0.4- to 1.1-μm spectral region. A milliammeter gives
an instantaneous indication of current from the cells, which can be interpreted in terms of incident energy flux. Another configuration has four
cells exposed to the upward hemisphere and four to the downward hemisphere, the difference in the output being integrated to give the net short-
wave radiation flux over the integration period.

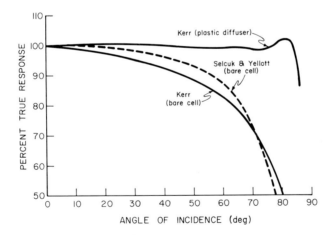

Fig. 4.22 Angular response of silicon solar cells with and without a plastic diffusing
cover, expressed in percent of the response according to the ideal cosine law (data from
Kerr *et al.*, 1967; and Selcuk and Yellott, 1962).

Fig. 4.23 An integrating pyranometer based on a combination of four silicon solar cells (courtesy of Science Associates).

4.2.3 Directly Integrating Pyranometers

Gunn–Bellani Integrating Spherical Pyranometer The importance of measuring the energy of solar radiation in biological systems in the field was emphasized by the English investigator P. A. Buxton in the middle 1920s, but an instrument with suitable reliability, accuracy, simplicity, robustness, and moderate expense did not exist. Buxton (1926), by a modification of a "radio integrator" designed by W. E. Wilson, con-

structed an integrating radiometer based on the multistep process of radiant energy being absorbed by a blackened surface enclosed in a spherical
glass bulb, the heat being transferred by conduction to a volatile liquid
(alcohol), where it is transformed into latent heat by evaporation of the
liquid. The distillation vapor is transferred to a condensation chamber
where it is recondensed, the heat of condensation being transferred by
conduction to the environment. The volume of liquid which condenses and
accumulates in the bottom of the tube is an indication of the amount of
energy originally absorbed by the blackened receiver. The instrument is
reset at the end of each period of measurements, for instance, each day, by
simply inverting the device and allowing the liquid to drain back into the
spherical receiver reservoir.

The design of the instrument has undergone a number of modifications in arriving at its present form. In Wilson's instrument the receiving
material was enclosed by a single thickness of glass which was freely exposed to the air, thereby causing the instrument to be very sensitive to
ambient air temperature and wind speed. Buxton (1926) greatly decreased
the sensitivity to the external environment by enclosing the receiving bulb
in an evacuated glass jacket, but he was still unable to obtain consistency
of reading among various instruments and the distillation efficiency was
only about 4% (Gunn *et al.*, 1945). Thus 96% of the heat loss was due to

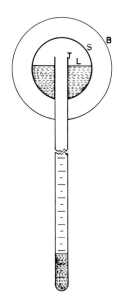

Fig. 4.24   Schematic diagram of the Gunn–Bellani integrating spherical radiometer.
S, copper sphere; T, glass distillation tube; L, distillation liquid; B, evacuated glass bulb.

nonevaporative processes and small variations in the amount of distillate collected. The consequences are that Buxton's model was both insensitive and inaccurate. In an effort to improve the performance, a second model, in which the distance between evaporator and condenser was shortened, was constructed by Buxton (1927), but it was apparently not tested in field conditions.

Gunn *et al.* (1945) incorporated Buxton's shortened diffusion route into a radiation integrator which was both more sensitive and more accurate than previous instruments of this general type had been. The design is shown schematically in Fig. 4.24. The receiver is a 5-cm-diameter copper sphere S which is coated on the outside with lamp black from a coal gas and benzene flame, mounted on a glass tube T, and enclosed in an evacuated glass bulb B. Copper assures efficient transfer of heat from the absorbing surface to the liquid. Gunn *et al.* considered that one of the reasons Buxton's instrument was so inefficient was that the vapor was forced to diffuse through some residual air in its movement from the evaporation chamber to the condensation chamber. They were careful, therefore, to obtain a vacuum of at least 10^{-5} mm Hg in their new models. Two types of volatile liquids—alcohol and water—are used, the choice depending on the sensitivity desired. In the water-filled instruments there was a tendency for droplets to collect on the sides of the condenser instead of joining the liquid in the graduated volume, but this difficulty was overcome by the addition of a small amount of soap to the water.

By comparing the integrated radiation values obtained from a number of these instruments with those from an Ångström pyrheliometer, Gunn *et al.* obtained some wildly erratic efficiency variations, the efficiency varying with both the radiant intensity and the particular instrument tested. Measured efficiencies varied from 0.0 to 107.9% when compared with the 100% efficiency assumed for the Ångström instrument. The mean efficiency of all measurements was perhaps 80%.

Various models of the instrument are available, each of which has its own features designed to increase the efficiency and repeatability of measurement. For instance, a Swiss model has the receiver covered with a gray metallic coating, while a German version uses a black glass sphere which is of a rough texture on the inside to promote increased evaporation efficiency.

Recent investigations have revealed an additional serious difficulty with at least some of the Gunn–Bellani-type instruments, namely, the existence of a temperature-dependent radiation threshold below which no distillation of the liquid occurs. Pereira (1959) first observed the effect and found the radiation threshold to be about 150 cal cm^{-2}/day for water-filled instruments in the tropics. A detailed investigation of the problem by Monteith and Szeicz (1960) showed that for a given ambient tempera-

ture T_0, there is a critical receiver temperature T_c which must be maintained or exceeded for distillation of the liquid to take place. Radiation
fluxes which are insufficient to maintain the required temperature difference $T_c - T_0$ between receiver and environment are not indicated by the
instrument. The critical value of radiation required to maintain the required temperature difference $T_c - T_0$ varies with T_0, as shown by Fig.
4.25. The radiation threshold is higher for water-filled instruments than for
alcohol-filled instruments. The effect is, furthermore, a function of the air
pressure inside the bulb. The data of Fig. 4.25 apply to instruments with
an internal pressure of 20–23 mm Hg, while the effect was not observed in
highly evacuated instruments. Finally, the calibration data, which have
been obtained with an uncollimated source in the laboratory do not necessarily apply to the combined direct-diffuse radiation field encountered in
realistic field applications.

Because of its basic design, the sensitivity of the Gunn-Bellani integrating radiometer is somewhat dependent on ambient air temperature,
wind speed, and amount of alcohol in the receiver chamber. For a typical
example, one spherical pyranometer from the Physikalisch–Meteorologisches Observatorium, Davos, Switzerland (calibration sheet, unpublished) shows a calibration factor $K = 8.2$ cal cm^{-2} for each centimeter of
condensate at an ambient temperature $T = 0°C$, with a variation from
$K = 8.7$ at $T = -10°C$ to $K = 7.9$ at $T = +30°C$. The temperature de-

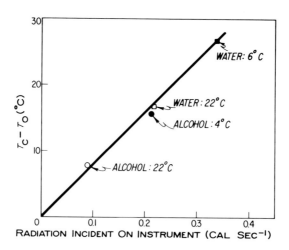

Fig. 4.25 Difference between the critical temperature for no distillation and the
ambient temperature of the environment as a function of incident flux for the Gunn–
Bellani integrating pyranometer.

pendence can be expressed as $K_T = K_0(1 - 0.0022T)$, where K_T is the value of K at temperature T and K_0 is its value at 0°C. The dependence on wind speed is less than 1% for wind speeds greater than about 1 m/sec. For lower wind speeds the error, caused by inefficient energy transfer to the environment, may exceed 10%, so mounting the instrument down inside a stand of plants should be avoided. The calibration factor increases with increasing volume of condensate in the tube, the increase being of the order of 2–4% in conditions of good ventilation. For low wind speed, the errors are greater.

Because of the relatively large mass of the receiver and liquid, the lag of the instrument is so great as to make long integration times mandatory. It is normally read once each day, either in early morning or late evening, as convenient, in which case the lag is of little consequence. However, for two hourly readings under changing radiation levels, a 10% error due to instrument lag can be expected.

4.2.4   Separation of Diffuse and Global Components

The total (global) solar radiation which is normally measured by a pyranometer consists of both the approximately parallel radiation which is transmitted directly through the atmosphere and the diffuse radiation from the sky. This latter component contains radiation which has been scattered from the solar beam on its initial downward traverse of the atmosphere and that which has been reflected from the surface and returned to the downward direction by the overlying atmosphere. It should be made clear that we refer here only to the solar radiation, as opposed to the downward terrestrial radiation which is emitted by the atmosphere itself.

For studies in atmospheric turbidity, ecology, and plant responses, and other problems, it is often desirable to measure the direct and diffuse components separately. One advantageous way of doing this is by the use of a solar-tracking pyrheliometer for determining the flux F_D of the direct radiation on a unit horizontal surface, together with a pyranometer for measuring the flux F of global radiation on a unit horizontal surface. Then the flux F_d of diffuse radiation on a unit horizontal surface is simply

$$F_d = F - F_D = F - F_p \cos \theta_0 \qquad (4.13)$$

where θ_0 is the zenith angle of the Sun and F_p is the flux of direct radiation on a unit surface normal to the solar beam. This is the method which is used at present by the National Weather Service.

Unfortunately this method requires an equatorial mount with active tracking of the Sun by the pyrheliometer. A simpler method for determining the direct and diffuse components separately, and one which is used frequently in different parts of the world, is by simultaneous measurements with two pyranometers, the first being exposed to the total global radiation and the second being shaded from the direct radiation but exposed to the diffuse radiation. In this case F_d is measured directly and F_D is given by

$$F_D = F - F_d \qquad (4.14)$$

Shading of the second pyranometer is either by a disk which is made to move with the Sun so as to always cast its shadow on the pyranometer, or by means of a shadow ring. Because of the trouble in keeping the shading disk in proper adjustment and the expense of an equatorial mount, the shadow ring is the more popular of the two devices. A photograph of a typical shadow-ring installation is shown in Fig. 4.26. The price paid for using the shadow ring is the necessity of introducing a correction for the part of the diffuse radiation which is cut off from the sensor by the shadow ring.

Fig. 4.26 Typical installation of a pyranometer and shadow ring for the measurement of diffuse sky radiation (courtesy of the Eppley Laboratories).

The geometry of the shadow ring installation is shown in Fig. 4.27. The plane of the shadow ring is set parallel to the equatorial plane by inclining the ring in the north–south plane by an angle ϕ from the zenith, ϕ being the latitude of the installation. Obviously the inclination is toward the south in the northern hemisphere and the north in the southern hemisphere. As the declination of the Sun changes with time the ring must be moved along an axis, which is oriented parallel to the Earth's axis, so as to always shade the pyranometer from the direct solar radiation. The frequency at which the position of the ring must be adjusted is a function of the size of the pyranometer, the width of the ring, and the rate of change of the solar declination. The only requirement is that the entire glass dome of the pyranometer be always in shadow. For most installations an adjustment of the position of the ring once each day is desirable, except perhaps near the solstices when the solar declination is changing least rapidly; one adjustment every 2 or 3 days is probably sufficient at those times. The adjustments can normally be made a part of the routine operation of the station.

The problem of correcting the pyranometer data for the sky radiation which is intercepted by the shadow ring may be approached either theoretically or experimentally, although neither of these approaches is entirely

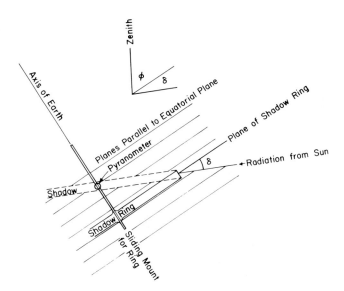

Fig. 4.27 Geometry of the shadow-ring installation for measurements of the flux of diffuse sky radiation.

satisfactory because of the variable nature of the intensity and its distribution over the dome of the sky. For the theoretical method, we see from Fig. 4.28 that the elemental solid angle $d\omega$ which is subtended by an elemental strip of the ring at the position of the pyranometer is

$$d\omega = (w \cos \delta/d^2)r \, dh \qquad (4.15)$$

where w is width of the ring, d is distance of the ring from the pyranometer, h is hour angle of the Sun, r is radius of the ring, and δ is declination of the Sun. Since $r = d \cos \delta$, we have

$$d\omega = (w/r) \cos^3 \delta \, dh \qquad (4.16)$$

The vertical component of the intensity I of diffuse radiation which is intercepted by the ring is $I \cos \theta$, where θ is the angle of the elemental area from the local zenith. Then the vertical component of the total flux of diffuse radiation intercepted by the ring is

$$\Delta F_d = \int_\Omega I \cos \theta \, d\omega = (w/r) \cos^3 \delta \int_{-h_0}^{h_0} I \cos \theta \, dh \qquad (4.17)$$

where h_0 is the value of h at sunset.

Since, by design, the zenith angles of the various elements of the ring

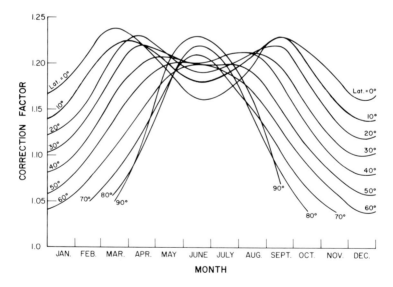

Fig. 4.28 Factors for correcting the measured flux of diffuse sky radiation for that part intercepted by a typical shadow ring at various latitudes as a function of time throughout the year.

are duplicated by the zenith angles of the Sun in its path across the sky we can write for the ring located at latitude ϕ:

$$\cos \theta = \sin \phi \sin \delta + \cos \phi \cos \delta \cos h \qquad (4.18)$$

or

$$\Delta F_{\mathrm{d}} = (w/r) \cos^3 \delta \int_{-h_0}^{h_0} I \left(\sin \phi \sin \delta + \cos \phi \cos \delta \cos h \right) dh \qquad (4.19)$$

In arriving at this expression we have assumed that the width of the ring is small compared to its radius, that the sensor is negligible in size compared to the other dimensions, and that there is no reflection of radiation from the inside of the ring.

Robinson and Stock (1964) have discussed other theoretical expressions for the shadow-ring correction. Of particular interest is a shadow-ring device which was developed by Robinson (1955), in which a set of 12 different shadow rings are provided. The particular ring to be used at a given time is dependent on declination of the Sun. The rings are sections of a sphere, and a special supporting member on the mount keeps the ring at a constant distance from the pyranometer. By varying the radius of curvature among the various strips it is possible to maintain a constant solid angle of sky which is occulted by the strip, regardless of the declination of the Sun.

The correction represented by Eq. (4.19) could be evaluated at any given time if the distribution of intensity through the region of the sky subtended by the ring were known. However, for practical purposes one is usually interested in the flux integrated over a finite time, such as an hour or a day. Since both the intensity and its distribution change markedly with Sun elevation, atmospheric turbidity, and cloudiness, Eq. (4.19) and similar expressions have limited applicability for a realistic situation.

Drummond (1956), Blackwell (1954), and others have simplified the theoretical analysis by assuming an isotropic intensity distribution over the sky. In that case Eq. (4.19) can be easily integrated with respect to time (hour angle) to obtain a correction factor for the shadow ring. Although, as pointed out by CSAGI (1958), this simplified analysis can provide a good guide as to the magnitude of the required correction, an experimental method for determining correction factors for a given location is probably the most satisfactory approach.

To obtain by experimental methods the correction for clear-sky conditions, the diffuse radiation as measured with the shadow ring in its normal position should be compared with that measured when the shadow ring is temporarily replaced by a shadow disk. The disk should be large enough to assure that the glass dome is completely shaded, and it should be held at such a distance that the angle subtended at the sensor by the disk is approximately 5°. For instance, for a 12-cm-diameter disk, its distance from the sensor should be about 137 cm. For cloudy-sky conditions (sky overcast, Sun invisible) the measurements with the shadow ring in place over the pyranometer should be compared with simultaneous measurements from a second pyranometer which is exposed to the complete global radiation. These procedures should be repeated enough times under various atmospheric conditions to give reliable correction factors.

In an extensive study of the shadow-ring correction, Drummond (1956, 1964) has developed correction factors for "average partly cloudy" skies which are applicable to the two models of the Eppley shadow rings. His tabulation was used to plot the curves of Fig. 4.28 in which the correction factor is given as a function of time for each 10° of latitude from the equator to the pole. Although the indicated time scale is for the northern hemisphere, the curves are equally applicable to the southern hemisphere if the time scale is shifted by 6 months. A 4.0% additive correction was incorporated into the data by Drummond to take account of the effects of a nonisotropic distribution of intensity over the sky. The corrections shown should be increased by about 0.02 for clear skies and decreased by 0.02 for overcast skies. It should be realized that these corrections are only approximations for a particular type of shadow ring and for general conditions. They should not be viewed as a substitute for corrections which are derived for a specific instrument at a given location.

4.2.5 Installation and Maintenance of Pyranometers

The details of installation and maintenance of pyranometers vary with the different types of instruments, and manufacturer's instructions should be closely followed. However, some general principles apply to all pyranometers.

The prime consideration in the installation of a pyranometer is that of proper exposure. Ideally, the sensing element for the hemispheric-type pyranometer should be above all surrounding obstacles such as buildings, trees, mountains, etc. so that the radiation can reach the instrument from the entire hemisphere of the sky. In practice, however, this ideal situation is not always realized, and some compromises, perhaps entailing corrections of the measurements, are necessary. The most serious obstructions are those which cut off the direct solar radiation from the instrument, so obstructions in the easterly and westerly directions are to be particularly avoided.

It is a simple matter by the use of a transit or theodolite to map on a diagram of azimuth versus elevation angles the outline of an obstacle, and determine the solid angle which it subtends at the instrument. If ϕ is azimuth angle and θ is zenith angle, then the solid angle subtended at the instrument by the obstacle is

$$\Omega = \int_{\phi_1}^{\phi_2} \int_{\theta_1}^{\pi/2} \sin \theta \, d\theta \, d\phi \qquad (4.20)$$

where the integration is taken over the projection of the obstacle above the horizon. If the obstacle were a blackbody, it would diminish the diffuse flux on a unit horizontal surface by an amount

$$\Delta F_{\rm d} = \int_{\phi_1}^{\phi_2} \int_{\theta_1}^{\pi/2} I(\theta, \phi) \, \cos \theta \, \sin \theta \, d\theta \, d\phi \qquad (4.21)$$

where $I(\theta, \phi)$ is the wavelength-integrated intensity of the skylight in the absence of the obstacle. However, a real obstacle has a finite reflectance, so in computing $F_{\rm d}$ from Eq. (4.21), $I(\theta, \phi)$ must be replaced by the intensity difference

$$\Delta I = I(\theta, \phi) - I'(\theta, \phi) \qquad (4.22)$$

where $I'(\theta, \phi)$ is the intensity of reflected radiation from the obstacle.

A realistic evaluation of ΔI is not so simple, however. Both I and I' are functions of position of the Sun and atmospheric conditions, and they have different wavelength distributions. The skylight is normally composed mainly of ultraviolet and visible radiation, whereas mineral-type surfaces such as concrete or stone reflect well at all wavelengths, with a slight increase toward the near infrared, and vegetative surfaces reflect little in the visible and ultraviolet but very strongly in the near infrared. Furthermore, the magnitude of the reflectance itself is quite different for a white rooftop from that for the side of a brick building or a distant mountain. Because of these complications it is preferable to choose a location for the pyranometer such that the error of measurement due to the obstruction is less than that introduced by other factors.

An estimation of the error likely to be introduced by a given obstacle may be illustrated by the following example. We assume an obstacle extending to 15° (0.262 rad) above the horizon through an azimuth range of 45° (0.785 rad), a skylight intensity distribution such that I corresponding to the direction of the obstacle is twice that of the average $\bar{I}$ over the sky, and that the diffuse radiation $F_{\rm d}$ contributes 15% of the total global flux F. Then by taking averages of the trigonometric functions in Eq. (4.21), we find the flux of diffuse radiation which is intercepted by the obstacle as

$$\Delta F_{\rm d} = (2\bar{I}) (0.129) (0.983) (0.262) (0.785) = 0.0522\bar{I} \qquad (4.23)$$

Since the flux of diffuse radiation from the whole sky is $F_{\rm d} = \pi \bar{I}$, and by our assumption $F_d = 0.15F$, we have the fraction of the total radiation intercepted by the obstacle as

$$\Delta F_d/F = (0.15)(0.0522)/3.14 = 0.00249 \qquad (4.24)$$

So far we have assumed the obstacle to be completely black, but for a real obstacle we can replace I in the above by ΔI of Eq. (4.22). Furthermore, in view of the usual errors in radiation measurements, a tolerance of up to 0.005 seems reasonable. Thus for the obstacle assumed, the prescribed conditions are met if the intensity of radiation reflected from the obstacle is within the range $0 \le I' \le 4\bar{I}$.

In general, the effect of an obstacle varies with position of the Sun, but unless the obstacle produces a fractional error of less than 0.005 for the worst case, appropriate corrections should be applied to the data, or a better location should be found for the instrument.

For practical purposes the pyranometer should be easily accessible, as the glass envelope must be kept free of dust or other deposit by frequent cleaning and any shadow rings or disks which are used require adjustment every 2 or 3 days for changes of declination of the Sun. The permissible distance between the pyranometer and recording device varies with types of both pyranometer and recorder; the manufacturers recommendations on the subject should be closely followed. An occasional check, at least once per year, of the resistance between the two conductors and between each conductor and the shield or conduit at (but disconnected from) the recorder will reveal any problems with deterioration of the insulation. Visual inspection, particularly for possible defects of the glass envelopes or the security of their mounting, should reveal any other problems with the physical aspects of the installation.

4.2.6   Calibration of Pyranometers

The calibration of both pyranometer and recorder should be checked at least once each year, and more often if feasible. Calibration of the recorder is a simple matter by the use of a standard potentiometer or other accurate source of dc voltage. Calibration of the pyranometer is best accomplished by comparison with a secondary standard instrument which has itself been calibrated against a primary standard. A recently calibrated operational-type instrument, preferably with temperature compensation, may be kept as a secondary standard for such routine calibrations, or if an interim replacement instrument is available the operational instrument may be sent to a central laboratory which is equipped for such calibrations.[*]

[*] Calibration facilities in the U.S. are maintained by the National Weather Service, Washington, D.C., and the Eppley Laboratories, Newport, R.I.

In using a secondary standard for field calibration of an operational instrument, the two are set up alongside each other on a clear day on which the radiation is not fluctuating rapidly, and a series of comparative measurements are made. Several hours record should be taken for the comparison. Mean hourly values F_1, F_2, . . . , F_n of the flux of radiation are computed from the millivolt output v_1', v_2', . . . , v_n' of the secondary standard over the n hours of measurement. If corresponding mean hourly values of the millivolt output from the operational instrument are v_1, v_2, . . . , v_n, then a series of values K_1, K_2, . . . , K_n of the calibration constant of the operational instrument are obtained from the relation

$$K_n = K'v_n'/v_n \qquad (4.25)$$

where K' is the calibration constant of the secondary standard. A final value for the operational instrument is the average of the n hourly values, or

$$K = (1/n) \sum K_n \qquad (4.26)$$

An alternate method of calibrating pyranometers is by the use of a well-calibrated auxiliary pyrheliometer for measuring the direct solar radiation and determining the decrease of signal from the pyranometer when it is shaded from the direct radiation by an appropriate size of opaque disk interposed between instrument and Sun. If F' is the radiative flux measured by the pyrheliometer, the flux F_D of direct radiation on a horizontal surface is $F_D = F' \cos \theta_0$, where θ_0 is zenith angle of the Sun. Then the calibration constant of the pyranometer is given by the simple relation

$$K = (v_2 - v)/F_D \qquad (4.27)$$

where v_1 and v_2 are the output signal from the pyranometer while it is shaded and unshaded, respectively. Best results are obtained by this method from a series of determinations made during a clear day when the Sun is high above the horizon. Probably the most critical feature in the method is to assure that the disk shades the entire glass envelope of the pyranometer and subtends approximately the same solid angle at the pyranometer as is encompassed by the field of view of the pyrheliometer. Only in that case is the pyranometer shielded from the same amount of radiation as that received by the pyrheliometer.

REFERENCES

Abbot, C. G., and Aldrich, L. B. (1916a). The pyranometer–An instrument for measuring sky radiation. *Smithson. Misc. Coll.* **66**(7), 9 p.

Abbot, C. G., and Aldrich, L. B. (1916b). On the use of the pyranometer. *Smithson. Misc. Coll.* **66**(11), 9 p.

Ångström, A. K. (1919). A new instrument for measuring sky radiation. *Mon. Weather Rev.* **47**, 795–797.

Ångström, A. K. (1924). Solar and terrestrial radiation. *Quart. J. Roy. Meteorol. Soc.* **50**, 121–126.

deBary, E. (1964). Influence of multiple scattering on the intensity and polarization of diffuse sky radiation. *Appl. Opt.* **3**, 1293–1303.

Blackwell, M. J. (1954). Five years continuous recording of total and diffuse solar radiation at Kew Observatory. Air Ministry Meteorol. Res. Comm., London, M.R.P. 895.

Budyko, M. I. (1963). "Atlas Teplovogo Balansa." Gidrometeorologicheskoe Isdatel'-skoe, Leningrad, U.S.S.R.

Buxton, P. A. (1926). The radiation integrator *in vacuo*, an instrument for the study of radiant heat received from the Sun. *J. Hyg.* **25**, 285–294.

Buxton, P. A. (1927). *Mem. London School of Hygiene and Tropical Medicine*, **1**, 22–36.

Callendar, H. L., and Fowler, A. (1906). The horizontal bolometer. *Proc. Roy. Soc. Ser. A* **77**, 15–16.

Chandrasekhar, S., and Elbert, D. D. (1954). The illumination and polarization of the sunlit sky on Rayleigh scattering. *Trans. Amer. Phil. Soc.* **44**, (6) 643–728.

Coulson, K. L. (1959). Characteristics of radiation emerging from the top of a Rayleigh atmosphere. *Planet. Space Sci.* **1**, 265–284.

Coulson, K. L. (1968). Effect of surface reflection on the angular and spectral distribution of skylight. *J. Atmos. Sci.* **25**, 759–770.

Coulson, K. L. (1969). Measurements of ultraviolet radiation in a polluted atmosphere. Sci. Rep. No. 1, Grant No. 5RO1 AP 00742-02, Univ. of California, Davis.

Coulson, K. L. (1971). On the solar radiation field in a polluted atmosphere. *J. Quant. Spectrosc. Radiat. Transfer* **11**, 739–755.

Coulson, K. L., Dave, J. V., and Sekera, Z. (1960). "Tables Related to Radiation Emerging from a Planetary Atmosphere with Rayleigh Scattering." Univ. of California Press, Berkeley.

Courvoisier, P. (1954). Theorie des bimetallaktinograph Robitzsch. Int. Radiation Conf., Rome (mimeographed).

Covert, R. N. (1925). Meteorological instruments and apparatus employed in the U.S. Weather Bureau. *J. Opt. Soc. Amer. Rev. Sci. Instrum.* **10**, 299–425.

CSAGI (1958). Radiation instruments and measurements, Part VI. "IGY Instruction Manual," pp. 371–466. Pergamon, Oxford.

Dave, J. V. (1964). Importance of higher order scattering in a molecular atmosphere. *J. Opt. Soc. Amer.* **54**, 307–315.

Deirmendjian, D. (1964). Scattering and polarization properties of water clouds and hazes in the visible and infrared. *Appl. Opt.* **3**, 187–196.

Deirmendjian, D., and Sekera, Z. (1954). Global radiation resulting from multiple scattering in a Rayleigh atmosphere. Tellus 4, 382–398.

Dirmhirn, I. (1958). Untersuchengen an Sternpyranometern. *Arch. Meteorol. Geophys. Bioklim. Ser. B* 9, 124–148.

Drummond, A. J. (1956). On the measurement of sky radiation. *Arch. Meteorol. Geophys. Bioklim. Ser. B* 7, 413–436.

Drummond, A. J. (1964). Comments on sky radiation measurements and corrections. *J. Appl. Meteorol.* 3, 810–811.

Drummond, A. J. (1965). Techniques for the measurement of solar and terrestrial radiation fluxes in plant biological research: a review with special reference to arid zones. *Proc. Montpiller Symp.*, UNESCO.

Drummond, A. J., and Roche, J. E. (1965). Corrections to be applied to measurements made with Eppley (and other) spectral radiometers when used with Schott colored glass filters. *J. Appl. Meteorol.* 4, 741–744.

Drummond, A. J., and Wade, H. A. (1969). Instrumentation for the measurement of solar ultraviolet radiation. *In* "The Biologic Effects of Ultraviolet Radiation," pp. 391–407. Pergamon, Oxford.

Drummond, A. J., Greer, H. W., and Roche, J. J. (1965). The measurement of the components of solar short-wave and terrestrial long-wave radiation. *Solar Energy* 9, 127–135.

Federer, C. A., and Tanner, C. B. (1965). A simple integrating pyranometer for measuring daily solar radiation. *J. Geophys. Res.* 70, 2301–2306.

Feigelson, E. M. (1964). "Radiation Processes in Stratiform Clouds" (in Russian). Science Publishing House, Moscow.

Feigelson, E. M., Malkevich, M. S., Kogan, S. Ya., Korontova, T. D., Glazova, K. S., and Kuznetzova, M. A. (1960). "Calculation of the Brightness of Light." (2 parts), (Translation from Russian). Consultants Bureau, New York.

Foster, N. B. (1951). A recording daylight illuminometer. *Illum. Eng. New York* 46, 59.

Fritz, S. (1955). Illuminance and luminance under overcast skies. *J. Opt. Soc. Amer.* 10, 820–825.

Greshchenko, Z. I. (1968). Relationship between the radiation regime and cloudiness. *Tr. Gl. Geofiz. Observ.* 223, 53–64.

Gunn, D. L., Kirk, R. L., and Waterhouse, J. A. H. (1945). An improved radiation integrator for biological use. *J. Exp. Biol.* 22, 1–7.

Hand, I. F. (1941). A summary of total solar and sky radiation measurements in the United States. *Mon. Weather. Rev.* 69, 95–125.

Haurwitz, B. (1948). Insolation in relation to cloud type. *J. Meteorol.* 5, 110–113.

Heinzpeter, H. (1952). Bericht uber neuere arbeiten zum solarimeter nach Gorczynski. *Z. Meteorol.* 6, (4), 118–121.

Kano, M. (1964). Effect of a turbid layer on radiation emerging from a planetary atmosphere. Doctoral Dissertation, Univ. of California, Los Angeles.

Kasten, F., Korb, G., Manier, G., and Moller, F. (1959). On the heat balance of the troposphere. Final Rep., AF CRC-TR-59-234, Gutenberg, Univ., Mainz, Germany.

Kerr, J. P., Thurtell, G. W., and Tanner, C. B. (1967). An integrating pyranometer for climatological observer stations and mesoscale networks. *J. Appl. Meteorol.* **6**, 688–694.

Kimball, H. H. (1914). Total radiation received on a horizontal surface from Sun and sky. *Mon. Weather. Rev.* **42**, 474–487.

Kimball, H. H., and Hand, I. F. (1922). Daylight illumination on horizontal, vertical, and sloping surfaces. *Mon. Weather. Rev.* **50**, 615–628.

Kimball, H. H., and Hobbs, H. E. (1923). A new form of thermoelectric recording pyrheliometer. *Mon. Weather. Rev.* **51**, 239–242.

Kipp and Zonen (1967). Private communication.

Kondratyev, K. Ya. (1965). "Actinometry." Hydrometeorological Publishing House, Leningrad (in Russian); translation NASA TT F-9712 (Nov. 1965).

Kondratyev, K. Ya. (1969). "Radiation in the Atmosphere." Academic Press, New York.

Lof, G. O. G., Duffie, J. A., and Smith, C. O. (1965). World distribution of solar radiation. *Sol. Energy* **10**, 27–37.

MacDonald, T. H. (1951). Some characteristics of the Eppley pyrheliometer. *Mon. Weather. Rev.* **79**, 153–159.

Marchgraber, R. M., and Drummond, A. J. (1960). A precision radiometer for the measurement of total radiation in selected spectral bands. Monograph No. 4, 10–12, Int. Union Geodesy Geophys., Paris.

Moll, W. J. H. (1923). A thermopile for measuring radiation. *Proc. Phys. Soc. London Sect. B,* **35**, 257–260.

Montieth, J. L., and Szeicz, G. (1960). The performance of a Gunn-Bellani radiation integrator. *Quart. J. Roy. Meteorol. Soc.* **86**, 91–94.

Moon, P., and Spencer, D. E. (1942). Illumination from a non-uniform sky. *Trans. Illum. Eng. Soc.* **37**, 707–726.

Morikofer, W., and Thams, C. (1937). Erfahrungen mit dem bimetallaktinographen Feuss–Robitzsch. *Meteorol. Z.* **54**, 360–371.

Neiburger, M. (1949). Reflection, absorption, and transmission of insolation by clouds. *J. Meteorol.* **6**, 98–104.

Pereira, H. C. (1959). Practical field instruments for estimation of radiation and evaporation. *Quart. J. Roy. Meteorol. Soc.* **85**, 253–261.

Plass, G. N., and Kattawar, G. W. (1968). Monte Carlo calculations of light scattering from clouds. *Appl. Opt.* **7**, 415–419.

Pochop, L. O., Shanklin, M. D., and Horner, D. A. (1968). Sky cover influence on total hemispheric radiation during daylight hours. *J. Appl. Meteorol.* **7,** 484–489.

Quinn, W. H., and Burt, W. V. (1968). Computation of incoming radiation over the equatorial Pacific. *J. Appl. Meteorol.* **7,** 490–498.

Robinson, N. (1955). An occulting device for shading the pyrheliometer from the direct radiation of the Sun. *Bull. Amer. Meteorol. Soc.* **36,** 32–34.

Robinson, N., and Stock, L. (1964). Sky radiation measurement and correction. *J. Appl. Meteorol.* **3,** 179–181.

Robitzsch, M. (1932). Uber den bimetallaktinographen Feuss–Robitzsch. *Gerlands Beit. Geophys.* **35,** 387–394.

Samuelson, R. E. (1965). Radiative transfer in a cloudy atmosphere. NASA TR R-215, Goddard Spaceflight Center, Greenbelt, Md.

Sekera, Z. (1956). Recent developments in the study of the polarization of skylight. *Advan. Geophys.* **3,** 43–104.

Selcuk, K., and Yellott, J. I. (1962). Measurement of direct, diffuse, and total solar radiation with silicon photovoltaic cells. *Sol. Energy* **6,** 155–163.

Thams, C. (1943). Uber die konstanz des eichfaktors beim bimetallaktinographen Fauss-Robitzsch. *Ann. Schweiz. Meteorol. Zentralanstalt,* **80,** 4–7.

Thekaekara, M. P. (1970). Proposed standard values of the solar constant and the solar spectrum. *J. Env. Sci.* **13,** 6–9.

Twomey, S., Howell, H. B., and Jacobowitz, H. (1967). Light scattering by cloud layers. *J. Atmos. Sci.* **24,** 70–79.

World Meteorological Organization (1965). Measurement of radiation and sunshine. *In* "Guide to Meteorological Instrument and Observing Practices," 2nd ed., Chap. 9. WMO-No. 8, TP 3 (Loose Leaf—1961).

Yanishevsky, Yu. D. (1957). "Actinometric Instruments and Methods of Observation." Hydrometeorological Publishing House, Leningrad (in Russian).

CHAPTER FIVE

Ultraviolet Radiation from the Sun and Sky

5.1 THE SOLAR ULTRAVIOLET

The Sun is the only source of ultraviolet radiation of any note for the Earth as far as the energy balance, biological activity, photochemical reactions, and other large scale phenomena are concerned. There are minute amounts of ultraviolet and shorter wavelengths of energy coming in from the stars, and of course some ultraviolet radiation is contained in the air glow and aurorae from the Earth's own atmosphere. Those radiations are studied mainly as indicators of processes taking place at the source of the radiation, and not for information on the radiation itself.

Studies of radiation from the Sun, on the other hand, have two important aspects. It is the most important indicator we have of the properties of the Sun and the physical and chemical reactions occurring on the Sun, but the radiation produces effects on the Earth which are of direct interest to Earth-borne dwellers. For instance, photochemical reactions due to ultraviolet radiation from the Sun produce ozone in the upper atmosphere which protects us from lethal doses of solar ultraviolet; longer wave ultraviolet which gets to the lower atmosphere promotes reactions for the production of photochemical smog; shorter wavelength ultraviolet and X radiations from the Sun ionize air in the upper levels of the atmosphere, producing the ionosphere which is so important for radio communications. These are examples from a host of effects resulting from ultraviolet, X-rays, and γ-rays emitted by the Sun.

5.1.1   Solar Ultraviolet Spectra

The term ultraviolet, often abbreviated as uv, is applied to the whole range of wavelengths from the lower end of the visible at approximately 0.4 μm down to X-rays beginning at about 0.03 μm. The exact boundaries are subject to interpretation among the various authors. The different parts of the uv spectrum are often designated as the near uv (approximately 0.4 to 0.3 μm), middle uv (0.3 to 0.2 μm), far uv (0.2 to 0.1 μm), and extreme uv (below 0.1 μm). The region below 0.2 μm is termed the vacuum uv, since vacuum techniques are necessary for measurements in that region. Another division is sometimes used, principally on the basis of biological effects, as uv–A (0.4 to 0.315 μm), uv–B (0.315 to 0.280 μm), and uv–C (wavelengths < 0.280 μm).

The ultraviolet solar spectrum is very rich in detail, due to numerous emission and absorption lines of the different elements in the solar chromosphere and corona. The most abundant of the elements is hydrogen, which is responsible for the very strong Lyman resonance line at 0.1216 μm, the broad continuum covering the range 0.0912 μm to about 0.0650 μm, and a number of lines between. The hydrogen series was photographed by T. Lyman in 1914. A similar but weaker continuum of helium occurs in the 0.05 μm region which also has strong resonance lines at 0.0304, 0.0537, and 0.0584 μm. Other strong emission lines are due to iron, silicon, oxygen, nitrogen, neon, carbon, and other elements (Hinteregger, 1970; Goldberg et al., 1968).

Spectra in the extreme ultraviolet are bright-line spectra, the radiation having come from the very high-temperature regions of the Sun. Longer wavelength spectra, however, are dark-line spectra, due to absorption in the solar atmosphere overlying the lower level and cooler regions principally responsible for the longer wavelength radiation. Because of its meteorological interest, several studies of solar spectra in the middle uv have been reported (e.g., Dunkleman and Skolnik, 1959; Kachalov and Yakovleva, 1962; Detwiler et al., 1961; Furukawa et al., 1967).

5.1.2   Ultraviolet Radiation in the Earth's Atmosphere

As is well known, ultraviolet radiation from the Sun causes important effects on the Earth. Wavelengths in the region 0.295–0.40 μm are partially transmitted by the atmosphere and, as mentioned above, cause biological effects of great importance. This aspect will be discussed further in the next section. Essentially all of the incident solar radiation at wavelengths below 0.295 μm is absorbed by the atmospheric gases, mainly the Hartley band of ozone (discovered by W. N. Hartley in 1891) in the 0.220–0.295 μm region and by oxygen in the Schumann continuum (V.

Schumann, 1893) between 0.1759 and 0.1450 μm, in the Schumann–
Runge bands between 0.2026 μm and the continuum, and in the 0.130–
0.104 μm region (Friedman, 1960). Absorption by ozone in the Hartley
band is so strong that, in spite of the fact that the total ozone content in
a vertical column would be a layer only about 3 mm thick at standard
conditions, only one out of 10^{40} photons in the 0.25 μm region reaches the
ground. Absorption coefficients for ozone have been tabulated by Vigroux
(1953), Ny and Choong (1932, 1933), Inn and Tanaka (1953), and others.
Shorter wavelengths are absorbed higher in the atmosphere, the depth of
penetration depending on wavelength and ionization properties of the
gases. The altitude at which the intensity is reduced to $1/e = 0.3679$, its
value at the outside of the atmosphere, as given by Friedman (1960), is
shown as a function of wavelength in Fig. 5.1.

Part of the radiation in the near ultraviolet (0.295 to 0.40 μm) is
absorbed by the Huggins bands of ozone (discovered by W. Huggins in
1890), mainly in the region λ < 0.32 μm, and part is back-scattered to
space, but a significant amount reaches the surface. Spectra of the near-uv
radiation which reaches the top of the atmosphere and that which was
measured at a height of 3660 m by Stair and Ellis (1968) on Mauna Loa,
Hawaii are shown in Fig. 5.2. The measurements were made by means of a
double-quartz-prism spectroradiometer in conjunction with a filter radio-

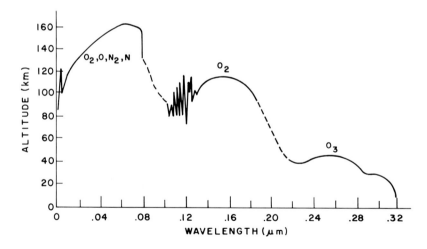

Fig. 5.1 Altitudes at which solar radiation intensity of various wavelengths is
reduced to $1/e = 0.3679$ of its value at the outside of the atmosphere. The materials
principally responsible for the attenuation are indicated in the applicable wavelength
regions (adapted from Friedman, 1960).

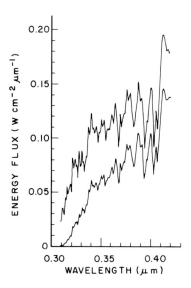

Fig. 5.2 Near-ultraviolet spectra of solar radiation at the top of the atmosphere and at a height of 3660 m on Mauna Loa, Hawaii (determined by Stair and Ellis, 1968).

meter which measures the flux in nine spectral regions in the 0.3 to 0.4 μm region. The extra-atmosphere curves were obtained by extrapolation to zero air mass on a plot of log D, D being instrument deflection, versus air mass in the usual manner. The authors estimate an accuracy of $\pm$ 5% for the curves.

5.1.3   Biological Effects of Ultraviolet Radiation

Biological effects of ultraviolet have been studied extensively, but many of the observed effects are not well understood. Ultraviolet therapy, either from sunlight or from artificial sources, has been used in many aspects, including the treatment of tuberculosis, healing of wounds, and in the prevention of spread of airborne diseases. Perhaps the most familar application is in the prevention and cure of rickets, an ailment of growing organisms caused by failure of lime salts to be deposited properly in the bones. Many applications of ultraviolet in industry and public health are based on its ability to destroy bacteria and viruses. Ultraviolet sources are sometimes installed in municipal water systems for bacterial control, and milk is sometimes pasteurized by ultraviolet irradiation. Unfortunately, the taste of milk has been found, at least in some cases, to suffer in this type of pasteurization. Experiments in Czechoslovakia have indicated an increase of growth rate and decrease of mortality of pigs, calves, and chickens by short exposures to ultraviolet light (Economic Commission for Europe, 1965). Most of these uses, of course, employ artificial sources of the radiation. Artificial sources are sometimes used in mines, factories, and submarines to take the place of natural sunlight, and the Germans provided ultraviolet treatment for their troops in Norway during World War II.

The wavelengths of radiation most effective in the prevention and cure of rickets is in the 0.270–0.313 μm range, with the maximum at 0.28 μm, as shown in Fig. 5.3 (Koller, 1965). Radiation in this range converts some of the sterols in the skin to vitamin D, which is necessary for the growth of healthy bones. The only part of solar radiation in the effective range which penetrates to the lower levels of the atmosphere is a small amount above 0.295 μm. It is interesting to speculate what would have happened to life on Earth if the Huggins band of ozone absorption were just slightly stronger, thereby preventing radiation in this range from reaching the surface.

An all too familiar effect of ultraviolet radiation is in the production of sunburn (erythema of the skin). Erythema ordinarily appears in 1 to 5 hr after exposure and lasts 1 to 3 days. The relative effectiveness of the various wavelengths in producing erythema, as given by Everett *et al.* (1969), is shown by the dashed curve of Fig. 5.3. This is quite a different distribution from that for antirachitic efficiency, the minimum at 0.28 μm

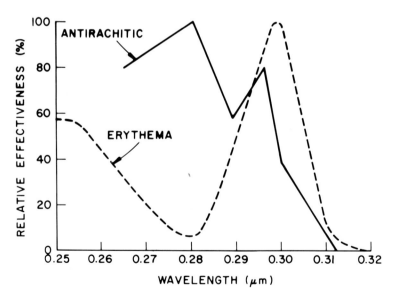

Fig. 5.3 Spectral distribution of the relative effectiveness of ultraviolet radiation for preventing rickets (solid curve) and in producing sunburn of the skin (dashed curve).

here coinciding with the maximum for vitamin D production. They show roughly comparable effectiveness, however, in the all important region of 0.295–0.310 μm.

Another consequence of exposure to ultraviolet is tanning of the skin. The production of tanning, which consists of the formation of the pigment melanin and a migration of melanin from lower layers of the skin, is most effective at ultraviolet wavelengths, but wavelengths as long as 0.65 μm produce the tanning reaction. Although tanning is the most obvious protection the body has to ultraviolet radiation, another protective mechanism is the thickening of the outer layer of the skin, which serves as a screen to short wavelengths of the ultraviolet. Of the large number of suntan preparations on the market, the most effective in protecting the skin from sunburn are those containing paraaminobenzoic acid, benzophenones, acrylonitriles, and tannic acid (Koller, 1965).

Ultraviolet radiation is known to be a major factor in causing skin cancer. Although the action spectrum for carcinogenesis is not known, the biologically most active range is 0.20–0.30 μm. The genetic material DNA is most susceptible to radiation at 0.26 μm, but effects are produced by longer wavelengths also. Information from the American Cancer Society shows 118,000 new cases of skin cancer each year in the U.S., mostly among the Caucasian population. The geographical distribution shows California has the highest rate of incidence, with 12,000 new cases per year. There is presently great concern regarding possible effects of supersonic transport aircraft on the ozone content of the atmosphere, with the attendant projected increase of ultraviolet radiation reaching the surface. Estimates indicate that a 5% decrease in ozone would result in 8,000 new cases of skin cancer annually in the U.S. A panel of the National Academy of Sciences has recommended the establishment of a network of stations at different latitudes and altitudes for measuring the incoming ultraviolet and establishing a baseline against which future changes could be evaluated (*Biomedical News*, April 1973).

Climatological measurements of ultraviolet radiation has received little emphasis in the U.S. and in most other Western countries. The observational data are few and sporadic, fluxes have generally been integrated over periods of a day or more, and there has been minimal standardization of the instrumentation. Hopefully the new types of instrumentation, particularly the uv fluxmeters for continuous recording of ultraviolet fluxes, will generate more interest in the subject. In contrast, considerable effort has been spent over the last decade or so in studying the ultraviolet climatology of Australia (Robertson, 1969) and of the territory of the Soviet Union. The results of a series of measurements at different locations in the Soviet Union (e.g., Vladivostok: Fedorets, 1966; Central

Asia: Sitnikova, 1966; Evpatoriya: Semenchenko, 1966; Tashkent: Generalov, 1967; Northern Regions: Galinin, 1962) have been combined with the pioneering work of Professor Belinskii on observations and atmospheric modeling in the very complete monograph Belinskii *et al.* (1968). A total of 31 maps of the geographical distribution of ultraviolet radiation in various wavelength intervals and time periods are included, as are discussions of ultraviolet effects and computations of ultraviolet flux based on Belinskii's transmission model. The geographical distribution is principally zonal in nature, as shown by the typical pattern of total flux for June at wavelengths < 0.40 μm in Fig. 5.4, although the obvious oceanic effect in the area of Eastern Siberia shows up in several of the results. A similar latitudinal distribution for the whole Earth was obtained by Schulze and Grafe (1969) from computations based on measurements of the global ultraviolet flux at a single station (Davos, Switzerland). These data, while valuable as preliminary estimates, should be further confirmed by many measurements of the ultraviolet fluxes in various locations and under different meteorological situations in order to provide a firm foundation for heliotherapy and other uses of natural ultraviolet radiation.

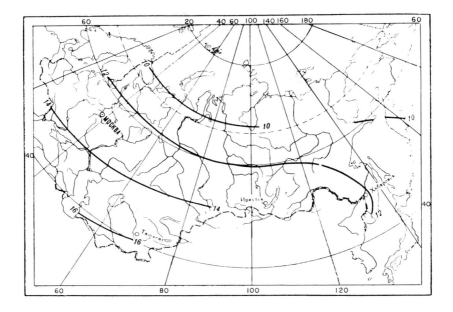

Fig. 5.4 Geographical distribution of the total flux of ultraviolet radiation on a horizontal surface (in units of kW hr/m²) as given for the territory of the Soviet Union (Belinskii *et al.*, 1968).

The role of the ultraviolet in producing photochemical air pollution is well documented, although the list of possible reactions is not completely known. One of the effects of air pollution on the ultraviolet itself is a significant decrease of the flux transmitted by the atmosphere for cases of even moderate pollution. Typical measurements of radiation flux in the 0.30 to 0.38 μm region reported by Stair and Nader (1967) for cases of clear and polluted conditions in Los Angeles are shown as a function of time in Fig. 5.5. For this and similar cases of photochemical pollution the global flux of ultraviolet radiation in the forenoon is less than half as much as it is for the case of no air pollution. Evidence indicates that the radiant energy is used in promoting photochemical reactions between oxides of nitrogen and hydrocarbons, both of which are plentiful during the forenoon. By noon or shortly thereafter, the supply of one or both of these reactive substances has been depleted, thereby slowing down the photochemical reactions and permitting a greater portion of the incident radiation to be transmitted by the atmosphere. Even in the case of a complete lack of

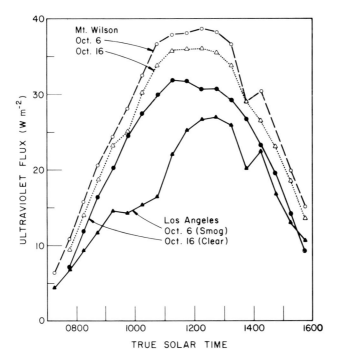

Fig. 5.5   Measured fluxes of ultraviolet radiation (λ = 0.30 to 0.38 μm) on a horizontal surface in downtown Los Angeles for cases of clear and polluted conditions, and at a location above the smog layer on nearby Mt. Wilson (data from Nader, 1967).

photochemical reactions, however, the flux would be decreased by simple absorption by the smoke, dust, and other particulates so numerous in a badly polluted atmosphere.

5.2   THE MEASUREMENT OF ULTRAVIOLET RADIATION

There are, in general, three main classes of detectors for measuring ultraviolet radiation, namely, physical, chemical, and biological. Examples of each of these will be discussed briefly below.

5.2.1   Biological Measurements of uv Doses

These measurements, used principally by the medical profession, are based usually on the amount of erythema or pigmentation of the skin in response to ultraviolet radiation, or on the effectiveness of such radiation in killing bacteria or viruses. A unit of ultraviolet dosage which produces a minimum perceptible erythema (MPE) is related to other amounts of exposure to the midsummer noonday Sun by the following rough approximations (see tabulation below).

Relative exposure (units of MPE)	1	2.5	5	10
Degree of erythema	MPE	Vivid, produces moderate tan	Painful burn	Blistering

For normal white skin, 1 MPE is about the equivalent of 2.5×10^5 ergs/cm^2 of radiation at 0.2967 μm wavelength (Koller, 1965).

Although such empirically derived indicators are useful in judging ultraviolet dosages, they suffer from being nonspecific and somewhat subjective, as well as from being inconvenient, having a long response time, and not being suitable for continuous recording.

5.2.2   Chemical Measurement of Ultraviolet Radiation

Chemical methods of measuring ultraviolet radiation were investigated in some detail in the 1920s and 1930s. Among the most promising methods at that time were the reactions of methylene blue in acetone and the uranyl acetate–oxalic acid reaction. The total uv dose was determined by a quantitative measurement of the reaction products. The main attractiveness of these and other chemical methods for measuring uv radiation is their simplicity and adaptability for a time integration of uv fluxes. They were never very satisfactory for routine measurements, however, because they were difficult to calibrate, often nonreproducible, temperature sensitive, and not well adapted to short-period determinations.

Since those early efforts, a number of workers have returned to chemical methods. Some of them are described below.

Dosimeter of I. G. Farben This instrument, which dates from the 1930's and was produced commercially in Germany, is based on the fact that fuchsin–leucosulfite changes from colorless to red on being exposed to ultraviolet radiation of wavelengths $\lambda < 0.31$ μm. Its spectral response is similar to that of erythema. The device consists of a small quartz tube filled with the solution and a green color filter wedge of 12 fields. The red color produced by exposure of the tube to ultraviolet radiation for periods of $\frac{1}{2}$ to 5 min is counteracted by viewing the tube through one of the fields of the green filter wedge. The specific field which produces a neutral gray color indicates the step on an empirical exposure scale. A thermometer was built into the unit to provide corrections for the temperature sensitivity of the reaction. According to Landsberg (1937), instruments manufactured after 1935 were stable, and different instruments gave comparable results.

Landsberg's Glass Dosimeter This device is a rod 1 mm diameter and 5 cm long made of a glass containing small amounts of vanadium and cerium compounds (Landsberg and Weyl, 1939). The unexposed rod has a greenish color which changes progressively to colorless and various shades of red on exposure to ultraviolet radiation. The action spectrum is in the 0.34–0.40 μm range, with a maximum sensitivity at about 0.36 μm. Evaluation of the color change was made by viewing the end of the rod through a green filter; an empirical scale of 9 steps was based on the colors observed. An integration of ultraviolet fluxes over about three summer days produced saturation, after which no further color change occurred.

Sherrod *et al.* (1950) exposed a series of 11 rods to different integrated fluxes of ultraviolet in an effort to provide standard colors as a calibration mechanism. They found, however, that the colors were not stable with time, and were forced to resort to a complicated scheme of exposing 12 rods simultaneously with different portions of the rods exposed for providing an autocalibration of the color changes. Unfortunately, these refinements reduced the very attractive feature of simplicity of the system and decreased the integration time from a possible 3 days to 1 hr. Furthermore, the accuracy attainable is not precisely known. According to Sherrod *et al.* (1950) it is "well within 20%," indicating relatively gross determinations which are hardly sufficient for modern scientific work.

uv Film Pyranometer The reaction in this device is the photoisomerization of the chemical σ-nitrobenzaldehyde (ONB) to σ-nitrobenzoic acid

(Pitts *et al.*, 1968). A transparent film is prepared by dispersing ONB in a solution of methyl methacrylate in benzene and allowing the mixture to evaporate. The residue is a transparent film 2.2 mm thick, which increases its optical density on exposure to ultraviolet radiation. The data readout is by means of an ordinary infrared densitometer. The reaction has a quantum efficiency of 0.50, is sensitive to wavelengths in the 0.30–0.41 μm range, and is linear with intensity. The exposure time for one flux determination is about 2 min in noonday sunlit conditions. A mechanism was devised for exposing a sequence of films for obtaining a semicontinuous record of ultraviolet fluxes.

Other Chemical Detectors for the uv The photolysis of nitrogen dioxide in nitrogen is an ultraviolet-sensitive reaction, and has been tested by Gordon (1966, 1967) for measurements of ultraviolet radiation through a volume. However, the method is cumbersome and not readily adaptable to continuous measurements.

Neuberger (1967) has used the optical degradation of Plexiglas (type G) as an indicator of ultraviolet radiation, an increase of optical density of the material being produced by radiation of λ < 0.345 μm. The amount of degradation is an approximately linear function of ultraviolet flux, but it is also a function of temperature of the material. By measuring the change of optical density of plates of 0.8 mm thickness and making appropriate temperature corrections, it is possible, according to the author, to achieve an overall accuracy in ultraviolet flux determinations of ± 15% or better.

The photochemical reaction of the cyanide of crystal violet was the basis of an instrument developed and used by Miyake (1949) for the measurement of ultraviolet radiation in Japan. The solution was sealed in a quartz ampule and exposed to ultraviolet for a period of 5 to 60 sec, while being shaken to mix the solution. The violet color which appeared was compared to that of 10 color standards representing different uv doses. The action spectrum of the material in combination with the solar spectrum in the uv gave a sensitivity between about 0.285 and 0.345 μm, with a maximum at 0.315 μm. The response was approximately linear and independent of the temperature of the solution.

Lothmar and Rust (1966) developed an instrument for field measurements of uv radiation utilizing the hardening of a coat of photoresist (e.g., Kodak-KPR) as an indicator of ultraviolet flux. Photoresist is used in making printed circuit boards in electronics. It produces a hard film on exposure to radiation in the 0.27–0.45 μm region, the thickness of the film being approximately proportional to the total flux of radiation. By means of a series of absorption and interference-type filters, the spectral sensitivity of the instrument could be confined to ranges of 0.4–0.28, 0.4–0.315, or 0.315–0.28 μm. A series of measurements showed the device to be useful for field measurements in medical and bioclimatic problems.

5.2.3 Measurements of Ultraviolet Radiation by Physical Methods

Most modern instruments for measuring solar ultraviolet radiation are based on physical principles, and many different types of uv radiometers have been developed. It is impossible to cover them all in a short discussion, but those considered to be of most interest in meteorology and aeronomy will be described below.

The types of sensors applicable to measurements of ultraviolet may be listed briefly as (1) radiometric devices (sensing the heating effect of radiation), (2) phototubes (including photomultipliers), (3) photovoltaic cells, (4) ionization chambers, and (5) photographic films. Photoconductive cells are not used much for ultraviolet sensing. The discovery of ultraviolet radiation by J. W. Ritter in 1801 was made by observing the blackening of silver chloride, which is the basis of the photographic method today, and the spectra utilized in the solar constant determinations by the Smithsonian Institution were obtained by use of a bolometer (radiometric method). However, modern technology has developed photovoltaic cells, phototubes, and ionization chambers to such a high degree of perfection that the radiometric method has declined in relative importance for ultraviolet measurements.

For convenience in the discussion, the various instruments will be divided according to the spectral interval in which they are most useful.

Instruments for the Vacuum Ultraviolet ($\lambda < 0.20$ μm) Instruments for the vacuum ultraviolet have seen very rapid development since the advent of the space program. Characteristics of the various detectors applicable to measurements in the region have been summarized by Dunkleman (1963) and by Friedman (1963). Spectra are normally obtained by spectrographs fitted with photographic or photoelectric sensors. The photographic plate is particularly good for obtaining spectral details, while rapid scans of the spectrum are best made with a photoelectric sensor of some type. A third method of getting low resolution spectra is by filter radiometry.

Photocathode surfaces for use in phototubes and photomultipliers for the vacuum ultraviolet have been developed from a number of different types of metals and semiconductor materials. Quantum efficiencies (ratio of number of photoelectrons emitted to number of incident quanta) of the order to 0.1 or more at wavelengths below the indicated low wavelength threshold are produced by CsI (0.22 μm), RbI (0.175 μm), KBr (0.155 μm),

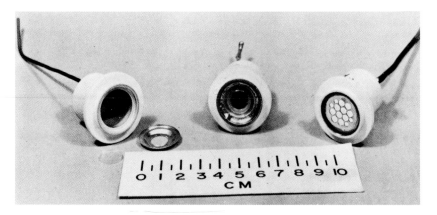

Fig. 5.6 Ion chambers for the measurement of radiation in the vacuum ultraviolet (after Dunkleman, 1963).

LiF (0.10 μm), and CuI (0.22 μm). Almost all pure metals have high work functions in the 0.14–0.10 μm region and can be used as windowless detectors in space applications.

Ion chambers are very useful detectors in the 0.10–0.16 μm region because of their high spectral selectivity. The basic construction is a small metal-lined cup, in the middle of which is mounted a pin electrode, as shown on the left of Fig. 5.6. The cavity is closed with a uv-transmitting sheet of material and filled with gas, and a bias voltage is impressed between the metallic lining and the pin electrode. Ultraviolet photons entering the chamber ionize the gas, resulting in a current through the system which is proportional to the number of incident photons. The spectral sensitivity is controlled by appropriate combinations of ionizing gas and window materials. Combinations for various spectral ranges, as given by Dunkleman (1963), are shown in Table 5.1.

A number of materials are suitable for making cutoff filters, the cutoff occurring on the short-wavelength side. The cutoff wavelengths (in μm) for various substances are the following: LiF: 0.105; CaF$_2$: 0.1225; strontium fluoride: 0.128; barium fluoride: 0.135; sapphire: 0.1425; pure silica: 0.160; crystalline quartz: 0.170.

Instruments for the Middle Ultraviolet ($0.2 \leq \lambda \leq 0.3$ μm)   Many different types of phototubes and photomultipliers are sensitive in the middle ultraviolet. Photocathodes of CsTe and RbTe not only have a strong response in that region, but they are also "solar blind" (have very low response in the visible and near ultraviolet, where the energy fluxes are

TABLE 5.1

Combinations of Ionizing Gas and Window Materials for Ionization Chambers
Sensitive in the Indicated Spectral Regions[a]

Type of gas	Window material	Spectral range (μm)
Ethylene oxide	LiF	0.105–0.118
Carbon disulfide	LiF	0.105–0.124
Acetone	CaF$_2$	0.123–0.129
Nitric oxide	CaF$_2$	0.123–0.135
Nitric oxide	LiF	0.105–0.135
Diethyl sulfide	BaF	0.135–0.148

[a] After Dunkleman, 1963.

much higher). Some of the more common commercial types of phototubes, particularly those with S1, S6, S13, and S19 responses,* are suitable for the middle uv if the envelope is made of quartz instead of glass. Photon counters, the receiving surface of which is one of a number of metallic elements, are sensitive in the middle uv, and they also have the feature of being solar blind.

By use of a conversion device, phototubes which are sensitive only to visible radiation may be used for ultraviolet measurements. This is accomplished by means of a fluorescent substance which absorbs ultraviolet energy and fluoresces at visible wavelengths. Zinc sulfide is suitable for wavelengths about 0.365 μm, magnesium tungstate for wavelengths of 0.285 μm, and willemite for wavelengths in the 0.2537 μm region. Sodium salicylate responds very well to ultraviolet throughout the 0.09 to 0.34 μm region and fluoresces at 0.443 μm, where many phototubes have a high response.

Instruments for the Near Ultraviolet ($0.3 \leq \lambda \leq 0.4$ μm)   Since this is the only region in which significant amounts of solar ultraviolet reach the ground, it is of great interest for practical purposes. As discussed above, it includes the uv which produces sunburn, tanning, and the production of vitamin D. Plants use some of the energy in the near uv for photosynthesis. Materials such as plastics and some paints are degraded by ultraviolet radiation, and the near uv is effective in producing photochemical products in air pollution. Because of such practical considerations, more effort has

* Response curve designations according to the Radio Tube Manufacturers Association.

been spent on instruments for this region than for those of shorter wavelengths.

Measurements by physical means are relatively simple in the near uv. There is enough energy for simple radiometric measurements, phototubes and photomultipliers are sensitive in the region, and there is no requirement for sophisticated vacuum techniques. In spite of these advantages, however, it has been only recently that reasonably satisfactory instruments for continuous measurements of ultraviolet have been developed for meteorological purposes. Those considered most applicable for continuous measurements, as well as some more specialized instruments, are described below.

Eppley Ultraviolet Pyranometer   This instrument is designed for continuous measurements of the global flux in the near-ultraviolet region. The sensor is a Weston selenium barrier-layer photoelectric cell with a sealed-in quartz window mounted inside a blackened tube beneath an interference filter and translucent quartz diffusing disk. The configuration of the device, as given by Drummond and Wade (1969), can be seen from Fig.5.7. The case is made of chrome-plated brass, sealed for exposure to the weather, and provided with a desiccant to prevent condensation of moisture inside

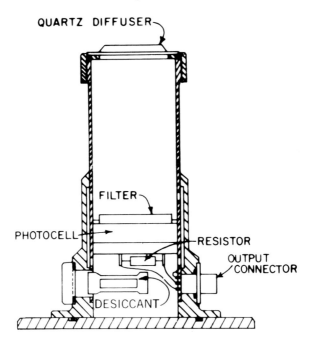

Fig. 5.7 Schematic diagram of the Eppley ultraviolet pyranometer (after Drummond and Wade, 1969).

the instrument. Other specifications as given by the manufacturer* are in the tabulation below.

Sensitivity	0.2 mV cal^{-1} cm^{-2} min^{-1}
Impedance	1000 Ω
Temperature dependence	-0.1% per 1°C from -40 to $+40$°C
Linearity	$\pm$ 2% from 0 to 0.1 cal cm^{-2} min^{-1}
Cosine response	to $\pm$ 2.5% for incident angles from 0° to 80°

The instrument is normally fitted with a filter which transmits over the spectral range of 0.295 to 0.385 μm. However, filters are available for narrower portions of the near-ultraviolet region. A quartz hemisphere may be installed over the diffusing disk for use in hostile environments. The broad-band device is calibrated by the manufacturer by reference to a calibrated thermopile installed alongside the pyranometer in the outdoor environment. The narrow-band calibration is by means of a standard lamp in the laboratory. Optional test equipment, consisting of three iodine–tungsten lamps with a regulated power supply, is available for checking the performance of the pyranometer in the field. Such a check once a month, in addition to daily inspections, is recommended.

For broad-band measurements, the recording equipment may be up to 980-m distance from the pyranometer, if the connecting cable is of 12 or 14 gauge wire and properly shielded. Shorter distances are recommended for the case of narrow-band measurements. According to Drummond and Wade (1969) measurement accuracy of better than $\pm$ 5% is possible with the instrument for broad-band measurements, with a precision of 1 to 2%. For narrow-band measurements the quoted accuracy is $\pm$ 10%.

CSIRO Ultraviolet Pyranometer The basis of this instrument is an ordinary network type of Kipp and Zonen pyranometer (see Chapter 4) fitted with a special ultraviolet filter (Collins, 1971). The outer glass hemisphere of the original pyranometer is replaced by a hemisphere made of type UG 11 Schott filter glass 8 mm in thickness, and the inner hemisphere by 2 mm thick type WG 10 filter glass. The infrared transmittance of the filter is less than 1/1000 of that for the ultraviolet. Unfortunately absorption of nonultraviolet radiation by the outer hemisphere causes its temperature to increase and radiation to be emitted to the sensor. The inner hemisphere is interposed between the two to shield the sensor from this long-wave energy.

Tests showed that an appreciable amount of heat from the filter was

* Eppley Laboratories, Inc., Newport, R.I. 02840.

transferred to the thermopile by conduction, thereby causing thermal drifts in the response. The effect was eliminated by chopping the incident radiation by means of a cam-actuated hemispheric cover used to alternately expose and shield the pyranometer at a rate of 1 cycle every 2 min. The amplitude of the resulting signal can be interpreted in terms of the ultraviolet flux incident on the pyranometer.

The UG 11 filter has a bell-shaped transmission curve covering the 0.29–0.38 μm spectral range with a maximum at about 0.33 μm. This spectral dependence coupled with a spectral dependence of incident flux on air mass makes the effective transmission, and thus instrument response, a function of air mass. A curve obtained by Collins (1971) showing effective transmittance versus air mass can be used to correct the results for the air mass effect.

Calibration of the device can be accomplished by the sun-and-shade method, with a pyrheliometer fitted with flat OG 11 and WG 10 filters. One difficulty is that wind effects in cooling of the outer hemisphere cause variations which prelude accurate integration of the signals over periods shorter than 1 hr. Collins (1971) reports successful operation of the instrument over several months, and concludes that it is suitable for routine ultraviolet measurements.

Volumetric Measurements of the Ultraviolet   For photochemical reactions in air pollution and for some biological processes, the direction of incidence of the radiation is immaterial, in which case the quantity of interest is the radiation through a volume instead of that through a plane surface. An interesting omnidirectional uv radiometer was devised by Lindh *et al.* (1964), by mounting ordinary barrier-layer selenium photovoltaic cells with appropriate filters on the six faces of a cube. The filters used were of Corning 7-37 polished glass with a peak transmission of 32% at 0.36 μm wavelength. The cells were connected by a network of variable resistors for taking account of differing sensitivities of the cells and providing a voltage signal which was proportional to the sums of the currents through the six cells, and thus to the total flux of radiation on the six faces of the cube.

The fact that the directional sensitivity of the individual cells was considerably different from the ideal cosine response was used to minimize the directional sensitivity of the composite. Two of the Corning 7-37 filters were mounted on each cell to increase the deviation from the cosine law. Data obtained by the authors indicated the response of the device to be independent of direction within ± 3.5%, which is certainly adequate for many practical purposes. The most serious deficiency of the instrument was a significant temperature sensitivity, which was not evaluated.

The basic configuration of the Lindh volumetric radiometer was adopted also for the instrument reported by Nader and White (1969). The sensors for this latter case were six of the Eppley ultraviolet pyranometers described above. The sensitivity of the Eppley pyranometers encompasses the 0.30–0.38 μm wavelength range, with a maximum at about 0.334 μm. In the cubical configuration, the sensitivities of the individual sensors were equalized by means of a series of 500-Ω resistors. The output signal from each of the sensors could be recorded separately as a measure of the radiation on each face of the cube, and the total output was a measure of the volumetric flux of ultraviolet radiation.

The signal from this volumetric radiometer was independent of direction to within 3.2%, which is close to that of the instrument of Lindh et al. (1964). The temperature sensitivity of the Nader and White device was presumably the same as that of the basic uv pyranometers (-0.1% per 1°C). Measurements showed the volumetric flux of ultraviolet radiation to exceed that through a horizontal surface by factors of 2.6 to 3.3 in polluted atmospheric conditions, and from 2.0 to 3.3 in a clean atmosphere.

REFERENCES

Belinskii, V. A., Garadzha, M. P., Mezhannaya, L. M., Nezbal, E. I. (1968). "Ultraviolet Radiation of the Sun and Sky." Moscow University Press, Moscow (in Russian).

Collins, B. G. (1971). A pyranometer for measuring ultraviolet radiation. *Aust. Meteorol. Mag.* **19**, 141–148.

Detwiler, C. R., Garrett, D. L., Purcell, J. D., and Tousey, R. (1961). The intensity distribution in the ultraviolet solar spectrum. *Ann. Geophys.* **17**, 263–272.

Drummond, A. J., and Wade, H. A. (1969). Instrumentation for the measurement of solar ultraviolet radiation. *In* "The Biologic Effects of Ultraviolet Radiation" (F. Urbach, ed.). Pergamon, Oxford.

Dunkleman, L. (1963). Ultraviolet photodetectors. Tech. Note D-1718, NASA Goddard Space Flight Center, Greenbelt, Md.

Dunkleman, L., and Scolnik, R. (1959). Solar spectral irradiance and vertical atmospheric attenuation in the visible and ultraviolet. *J. Opt. Soc. Amer.* **49**, 365–367.

Economic Commission for Europe, Committee on Electric Power (1965). The use of ultraviolet rays in agriculture. Working Paper No. 87 (mimeographed).

Everett, M. A., Sayre, R. M., and Olson, R. L. (1969). Physiologic response of human skin to ultraviolet light. *In* "The Biologic Effects of Ultraviolet Radiation" (F. Urbach, ed.). Pergamon, Oxford.

Fedorets, B. A. (1966). Hygienic features of ultraviolet radiation in Vladivostok. *Gig. Sanit.* **31**, 82–85.

Friedman, H. (1960). The Sun's ionizing radiations. *In* "Physics of the Upper Atmosphere" (J. A. Ratcliffe, ed.). Academic Press, New York.

Friedman, H. (1963). Ultraviolet and X-rays from the Sun. *Ann. Rev. Astron. Astrophys.* **1**, 59–96.

Furukawa, P. M., Haggenson, P. L., and Scharberg, M. J. (1967). A composite, high-resolution solar spectrum from 2000 to 3600 A. Tech. Note. No. 26, National Center for Atmospheric Research, Boulder, Colorado.

Galanin, N. F. (1962). Ultraviolet climate of the northern regions of the U.S.S.R. *Leit. TSR Mosklu Akad. Geol. Geogr. Inst., Moksliniai Pranesimai,* **13**, 257–265.

Generalov, A. A. (1967). Hygienic evaluation of natural UV radiation in Tashkent. *Gig. Sanit.* **32**, 103–105.

Goldberg, L., Noyes, R. W., Parkinson, W. H., Reeves, E. M., and Withbroe, G. L. (1968). Ultraviolet solar images from space. *Science* **162**, 95–99.

Gordon, R. J. (1966). Photochemical measurements of ultraviolet sunlight. Paper No. 66-38, Meeting of Nat. Air Pollution Assoc., June 20–24, 1966. San Francisco, California.

Gordon, R. J. (1967). Photochemical measurements. *In* "Pilot Study of Ultraviolet Radiation in Los Angeles October 1965" (J. S. Nader, ed.). National Center for Air Pollution Control, Cincinnati, Ohio.

Hinteregger, H. E. (1970). The extreme ultraviolet solar spectrum and its variation during a solar cycle. *Ann. Geophys.* **26**, 547–554.

Inn, E. C. Y., and Tanaka, Y. (1953). Absorption coefficients of ozone in the ultraviolet and visible regions. *J. Opt. Soc. Amer.* **43**, 870–873.

Kachalov, V. P., and Yakovleva, A. V. (1962). The ultraviolet spectrum in the region 2470-3100Å. *Izv. Akad. Nauk, S.S.S.R. Krym. Astrophys. Obs.* **27**, 5–43 (in Russian).

Koller, L. R. (1965). "Ultraviolet Radiation." Wiley, New York.

Landsberg, H. (1937). The ultraviolet dosimeter. *Bull. Amer. Meteorol. Soc.* **18**, 161–167.

Landsberg, H., and Weyl, W. (1939). Measurements of ultraviolet sums with photosensitive glass. *Bull. Amer. Meteorol. Soc.* **20**, 254–256.

Lindh, K. G., Buchberg, H., and Wilson, K. W. (1964). Omni–directional ultraviolet radiometer. *Sol. Energy* **8**, 112–116.

Lothmar, R., and Rust, E. (1966). Die messung ultravioletter strahlung mit einem photoresist als empfanger: eine feldmethode zur bestimmung der UV–Strahlung von sonne and himmel. *Arch. Meteorol. Geophys. Bioklim. Ser. B* **14**, 360–383.

Miyake, Y. (1949). A new chemical method for measuring the ultraviolet ray. *Geophys. Mag.* **19**, 95–99.

Nader, J. S., ed. (1967). "Pilot Study of Ultraviolet Radiation in Los Angeles October, 1965." National Center for Air Pollution Control, Cincinnati, Ohio.

Nader, J. S., and White, N. (1969). Volumetric measurement of ultraviolet energy in an urban atmosphere. *Env. Sci. Tech.* **3**, 848–854.

Neuberger, H. (1967). Photosensitive plastic measurements. *In* "Pilot Study of Ultraviolet Radiation in Los Angeles October, 1965" (J. S. Nader, ed.). National Center for Air Pollution Control, Cincinnati, Ohio.

Ny, T. Z., and Choong, S. P. (1932). L'absorption de la lumiere par l'ozone entre 3050 et 3400 A. *Comp. Rend. Acad. Sci., Paris* **195**, 309.

Ny, T. Z., and Choong, S. P. (1933). L'absorption de la lumiere par l'ozone entre 3050 et 2150 A. *Comp. Rend. Acad. Sci., Paris* **196**, 916.

Pitts, J. N., Cowell, G. W., and Burley, D. R. (1968). Film actinometer for measurement of solar ultraviolet radiation intensities in urban atmospheres. *Env. Sci. Tech.* **2**, 435–437.

Robertson, D. F. (1969). Correlation of observed ultraviolet exposure and skin-cancer incidence in the population in Queensland and New Guinea. *In* "The Biologic Effects of Ultraviolet Light" (F. Urbach, ed.). Pergamon, Oxford.

Schulze, R., and Grafe, K. (1969). Consideration of sky ultraviolet radiation in the measurement of solar ultraviolet radiation, *In* "The Biologic Effects of Ultraviolet Radiation" (F. Urbach, ed.). Pergamon, Oxford.

Semenchenko, B. A. (1966). Direct ultraviolet radiation in Yevpatoriya. *Moscow University, Vestnik, Ser. 5, Geog.* **21**, 104–107.

Sherrod, J., Neuberger, H., and Yerg, D. (1950). New standards for Landsberg's glass rod method of integrating ultraviolet radiation. *Trans. Amer. Geophys. Union*, **31**, 696–698.

Sitnikova, M. V. (1966). Ultraviolet radiation on the territory of Central Asia. *Geliotekhnika*, **2**, 35–37.

Stair, R. and Nader, J. S. (1967). Filter-phototube measurements. *In* "Pilot Study of Ultraviolet Radiation in Los Angeles October 1965" (J. Nader, ed.). National Center for Air Pollution Control, Cincinatti, Ohio.

Stair, R., and Ellis, H. T. (1968). The solar constant based on new spectral irradiance data from 310 to 530 nanometers. *J. Appl. Meteorol.* **7**, 635–644.

Vigroux, E. (1953). Contribution a l'etude experimentale de l'absorption de l'ozone. Doctoral thesis, Fac. Sci., Univ. of Paris.

CHAPTER SIX

Illumination

6.1 DEFINITIONS AND UNITS

Radiation in the visible region of the spectrum is often evaluated with respect to its capability of evoking the visual sensation of brightness. A whole system of quantities and units, the photometric (or photovisual) system, has been built on this visual sensation of brightness for the normal light-adapted human eye. A considerable amount of confusion has resulted between the photometric system and the radiometric system of units, the radiometric system being entirely in terms of the physical energy of radiation and the photometric in terms of the effects produced by that energy. Although it is not possible in general to make a rigorous conversion between the two systems, the main physical concepts of intensity, flux, power, etc. are common to both. Furthermore, on the basis of experiments by many investigators to determine the relative effectiveness of radiant energy of a given wavelength to evoke the sensation of brightness, the Commission Internationale de L'Eclarage (C. I. E.) meeting in 1924 recommended, and later adopted (1931) a standard of sensitivity versus wavelength of the normal light-adapted human eye (standard observer in C. I. E. terminology). This sensitivity (also called luminosity) curve is shown in Fig. 6.1; the complete tabulation is included as Appendix B. As will be discussed later, it is possible to use the luminosity curve and a quantity called "monochromatic luminous efficiency" to convert from the photometric to the radiometric system for light of a known spectral distribution.

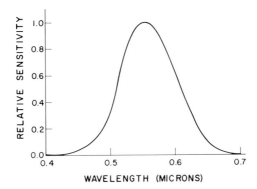

Fig. 6.1   Relative sensitivity (or luminosity) vs wavelength for the normal light-adapted human eye (C.I.E. standard observer, 1931).

Several different types of light sources have been used as photometric standards in different countries. The earliest standard in the United States and Great Britain was a spermaceti candle weighing $\frac{1}{6}$ pound and burning 120 grains/hr with a flame 45 mm high. Stearin candles were the standard in France, and German candles were of paraffin.

Because of difficulties of maintaining actual candles as standards, they were eventually replaced by standard lamps of the following types (Viezee, 1960):

Great Britain—Harcourt pentane lamp (1 pentane candle = 10 international candles)

France—Carcel lamp burning purified rapeseed oil (1 carcel unit = 9.6 international candles)

Germany—Hefner amylacetate lamp (1 hefner unit = 0.9 international candles)

United States—Carbon-filament incandescent lamp (introduced in 1909; defined both American and international candles)

A new unit of luminous intensity was introduced in the United States and Great Britain in 1948. The unit has been known as the new candle, the candela, and simply the candle. It is the luminous intensity in the direction normal to a $1/60$ cm^2 blackbody area at the solidification temperature of platinum (2042°K). The standard intensity of 1 cm^2 of the blackbody is thus 60 (new) candles, and it was found to be the equivalent of 58.84 international candles. Therefore the international candle has the value 1.019 (new) candles. For convenience, we will hereafter drop the designation "new" candle, and the unit candle will be that given by the blackbody standard at a temperature of 2042°K.

We are now in position to define the various quantities and units of the photometric system and compare the units to those of the radiometric system. The symbols are listed in the following tabulation.

A = area
t = time
ω = solid angle
γ = angle with respect to the surface normal
λ = wavelength
Q = energy evaluated with respect to its ability to evoke the sensation of brightness

L = luminous flux
K = luminous efficiency
f_λ = luminosity (eye sensitivity) function
B = luminous intensity
F = radiometric flux

$$\text{Luminous intensity:} \quad B = \frac{dQ}{d\omega \, dt} \tag{6.1}$$

The standard unit is the candle (or candlea). Luminous intensity, often called candle-power for obvious reasons, may be thought of as the solid angular density of luminous flux. One candle is equal to 1 lumen/steradian.

$$\text{Luminous flux:} \quad L = \frac{dQ}{dA \, dt} \tag{6.2}$$

The standard unit, the lumen, is the luminous flux emitted in all directions by a blackbody of surface area $(1/60 \, \pi)$ cm^2 at a temperature of 2042°K. The older definition of lumen as the luminous energy passing in unit time through a unit area at unit distance from a unit point source was rendered obsolete in 1948, when the new standard source was adopted. Luminous flux is used to express the rate of luminous energy output (luminous power) from a source, which may be either small or extended.

$$\text{Luminance (or brightness)} = \frac{dQ}{dA \cos \theta \, dw \, dt} \tag{6.3}$$

The term luminance corresponds to radiance in radiometric terminology, and connotes emitted luminous energy. It is used to characterize extended sources, and can be measured in luminous intensity per unit projected area. For instance the luminance of an ordinary fluorescent lamp is 3 to 4 candles/in^2 of tube surface. The units used for luminance are given in the following tabulation.

stilb	1 candle cm^{-2}
apostilb	1 candle m^{-2}
foot-candle	1 candle ft^{-2}
lambert	$1/\pi$ stilb
meter-lambert	$1/\pi$ apostilb
foot-lambert	$1/\pi$ foot-candle

The term brightness has the same connotation as luminance, but brightness is estimated visually. For very low light levels the eye loses some of its ability to sense brightness (the Purkinje effect), in which case brightness differs from luminance. If the physical phenomenon is under discussion, the correct term is luminance.

$$\text{Illuminance (or illumination)} \; = \; dL/dA \tag{6.4}$$

Illuminance corresponds to irradiance in the radiometric terminology and connotes incident energy, usually energy on a surface. Units of illuminance are given in the following tabulation.

lux	1 lumen m^{-2}
phot	1 lumen cm^{-2}
foot-candle	1 lumen ft^{-2}

$$\text{Luminous efficiency:} \quad K_\lambda \; = \; L_\lambda/f_\lambda \tag{6.5}$$

The conversion between the photometric and radiometric systems of units is accomplished by the use of K_λ and the luminosity function f_λ, the curve for which was shown in Fig. 6.1. The conversion is accomplished by the relation (Condon and Odishaw, 1958)

$$1 \; W \; = \; f_\lambda \; \text{lumens} \tag{6.6}$$

The maximum value of K_λ is $685f$ lumens W^{-1}, which occurs at $\lambda = 0.555$ μm. For the total (wavelength-integrated) luminous flux we have

$$K \; = \; \int_0^\infty 685 \, f_\lambda F_\lambda \, d\lambda \bigg/ \int_0^\infty F_\lambda \, d\lambda \tag{6.7}$$

where F_λ is the monochromatic radiant energy flux per unit area per unit time. For a given spectral energy distribution Eq. (6.7) can be evaluated, thereby yielding a unique value of K applicable to the given field of radiation. In particular, for a blackbody radiation F_λ can be computed as a

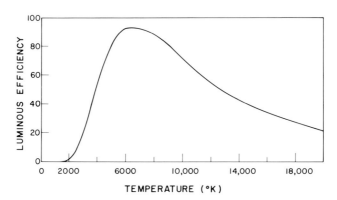

Fig. 6.2 Variation of luminous efficiency with temperature for a blackbody (units, lumens W⁻¹).

function of wavelength at a given temperature by Planck's law, from whence it is possible to determine K as a function of temperature for blackbody radiation. The resulting curve of K vs T for blackbody sources is shown in Fig. 6.2 (Viezee, 1960). It is interesting that luminous efficiency is a maximum at approximately the temperature of the Sun (6000°K), and it falls off rapidly at lower temperatures. The efficiency at the usual temperature of the tungsten filament of an incandescent lamp (approximately 2900°K) is only about 20% that at 6000°K, a fact which clearly shows the desirability of obtaining higher temperature lamps for use in providing illumination.

6.2 MEASUREMENT OF PHOTOMETRIC QUANTITIES

Requirements for measurements of photometric quantities are normally associated with either the characterization of sources of illumination (luminous intensity for a small area or point source and luminance for an extended source) or a definition of the illuminance (luminous flux or illumination) on a surface. The two requirements will be discussed separately.

For a characterization of sources of illumination, one is concerned mainly with the total amount of luminous energy emitted in a given direction, although its spectral distribution is of interest in cases in which color is important. The ordinary sources of illumination, such as incandescent lamps, fluorescent lamps, gas-filled lamps, etc., are not readily characterized from first principles, but must be referred to a standard source. As mentioned above, the standard source now used in the United States and Great Britain is a $1/60$-cm^2 blackbody at the solidification temperature of platinum ($2042°$K), while the standards for most other countries are calibrated tungsten lamps. These primary standards are maintained in national standards laboratories, and are used to calibrate secondary standard lamps for distribution to other laboratories. These secondary standards, in turn, are used to calibrate light sources for operational purposes.

6.2.1   Visual Photometers

Bouguer Photometer   A number of different devices have been developed for comparing an operational light source with a standard. The earliest known photometer was that of Pierre Bouguer (1698–1758) described in 1729. It consisted of two screens, in each of which was a small hole covered with translucent paper. The screens were set up vertically, one on each side of an opaque vertical partition. The standard and operational light sources were placed on opposite sides of the partition, and thus each lamp illuminated one of the screens. By adjusting the distance between lamps and screens, the brightness of the two holes in the screens could be made the same, in which case the relative intensities of the lamps could be calculated from the inverse square law.

Rumford Photometer   The scientist and mathematician J. H. Lambert (1728–1777) devised in 1760 a photometer for comparing two light sources by the shadows cast by opaque objects. Since a device of the same design was used more extensively by Count Rumford, it is known as the Rumford photometer. It consisted of a vertical screen with a matte white surface, in front of which were placed two vertical cylindrical sticks separated by a partition. The lamps to be compared were positioned so that the shadows of the two sticks fell next to each other on the screen. By varying the lamp-to-screen distances, the shadows could be made of equal intensity, thereby permitting calculation of relative intensities of the lamps by the inverse square law. A later variation utilized a second screen with an aperture cut in it to replace the stick, in which case the two images of the aperture on the screen were brought to equal brightness. These types of photometers were capable of yielding accuracies of ± 8 to $\pm 10\%$.

Bunsen Photometer   A photometer of early design which is still popular is the Bunsen or "grease-spot" photometer. A simplified diagram of the device is shown in Fig. 6.3. The movable screen S, made of white paper or parchment, has a spot which is made translucent by grease or oil in its center. The position of the screen between the two light sources L_1 and L_2 is adjusted so that the spot is made to disappear, at which time the illumi-

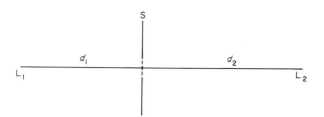

Fig. 6.3 Simplified diagram of the Bunsen or "grease-spot" photometer.

nation is the same on the two sides of the screen. Then the ratio of intensities I_1 and I_2 of lamps L_1 and L_2 is given by

$$I_1/I_2 = d_1^2/d_2^2 \tag{6.8}$$

where d_1 and d_2 are the lamp to screen distances, respectively.

In actual operation of the Bunsen photometer, the grease spot cannot be made to entirely disappear because of its somewhat different character from that of the rest of the screen. In order to minimize this effect, most present models of the instrument have the sheet mounted in a movable box which is fitted with mirrors to allow both sides of the screen to be viewed simultaneously. In this configuration the Bunsen photometer is capable of yielding relative intensities to perhaps $\pm$ 4 to $\pm$ 5% accuracy.

Lummer–Brodhun Photometer Probably the most accurate visual photometer is the Lummer–Brodhun photometer. A schematic diagram of this device is shown in Fig.6.4. As in the previous case, the movable screen S may be positioned between the light sources L_1 and L_2 for changing the intensity of light on the two sides of the screen, but the unique feature of this instrument is the optical system for viewing the two sides simultaneously. This consists of mirrors M_1 and M_2 which receive light from the respective sides of the screen and reflect it to the prism assembly P_1 and P_2. Both prisms are right angle prisms. Prism P_1 has all faces plane, but the hypotenuse face of P_2 is spherical except for a plane spot in the middle. This plane spot is attached with optical cement to the hypotenuse face of P_1. Light rays from M_2 which enter prism P_2 and strike the spherical surface are totally reflected thereon, whereas those which strike the cemented plane surface are transmitted and eventually end up at the viewing eyepiece E. Rays from M_1 which enter prism P_1 and strike the cemented spot are transmitted and lost, while those which strike outside of the spot are reflected to the eyepiece. Thus the field viewed in the eyepiece consists of a central ellipse lighted from lamp L_2 which is surrounded by an area lighted by L_1. Only when the light intensities on the two sides of the screen

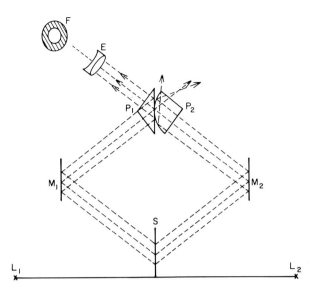

Fig. 6.4 Schematic diagram of the Lummer–Brodhun photometer. L_1 and L_2, light sources; S, matte-white opaque screen; M_1 and M_2, mirrors; P_1 and P_2, optical prisms; E, viewing eyepiece; F, intensity field as seen in eyepiece.

are the same does the spot disappear and the viewed field become of uniform intensity. This condition is achieved by proper positioning of the screen between the light sources, as in other types of photometers.

A variation of the Lummer–Brodhun photometer is designed for equalizing contrast instead of intensity, as is usually done. This contrast photometer has the contact area of the two prisms modified so that the field is split in two. The two sources each illuminate a trapezoidal area, one in each half of the field. In each case the intensity within the trapezoid is reduced about 8% by a thin glass plate inserted in the beam. The relative intensities of the beams are then adjusted by movement of the screen to give equal contrasts between the trapezoidal areas and their backgrounds in the two halves of the field. The eye is better able to judge contrasts of intensities themselves, a fact which explains the high sensitivity attainable with the contrast photometer.

Martens Polarization Photometer Of the several photometers utilizing the polarization of light, the only one which appears to have survived for modern use is that of Martens developed about 1900. For a description see Walsh (1958). In this instrument, the light streams from two comparison surfaces are polarized and then analyzed by a rotatable polarizing prism before being presented to the eye. The light to be measured is normally reflected from a highly reflecting plate and made to illuminate one-half of the viewed field. The other half of the field is an opal glass illuminated by a reference lamp. The light from each of the two surfaces is polarized by a Wollaston prism, but the planes of polarization of the two are at right angles to each other. By inserting a Nicol prism in the oppositely polarized streams and rotating the prism, it is possible to equalize the brightness of the halves of the field. Then the illumination L of the reflecting surface is given by $L = k \tan^2 \alpha$, where α is the angle of the Nicol from its position of maximum brightness of the reference surface, and k is a calibration factor of the instrument obtained from a secondary standard of luminous intensity.

In using the Marten's polarization photometer, the light from the comparison surfaces must be free of polarization. Since completely un-polarized light is very difficult to obtain, this residual polarization is a source of significant error in the measurements. Other errors are introduced by inexact alignment of the optical axes of the Nicol and Wollaston prisms and by errors in reading the angle of rotation. The result of all of these errors is to decrease the accuracy of measurement attainable with this instrument below that of an instrument such as the Lummer–Brodhun photometer.

Flicker Photometers   A difference of color between the light sources being compared introduces a great deal of difficulty in the use of visual photo-meters. In the Lummer–Brodhun photometer, for instance, the central spot of the field cannot be made to disappear in the case of sources of different color, as the color difference will show up between spot and back-ground. Various methods of making photometric measurements in such a case have been devised (Walsh, 1958), the most successful being that of the flicker photometer. The basis of this instrument is that equality of illumination is indicated by the disappearance of a flicker sensation as the two lighted areas are presented alternately to the eye. The optimum flicker frequency is dependent on the brightness of the fields, the difference in color of the sources, and the observer himself, but it should be as low as possible to produce the flicker sensation. At high speeds the flicker dis-appears regardless of the characteristics of the light sources. Different types of flicker photometers, including the Bechstein, the Ives–Brady, and the Guild models, have been described in detail by Walsh (1958).

Other types of visual photometers which have received some popu-larity are several models of the Weber photometer, and various photometers utilizing rotating sector disks (e.g., the Napoli–Abney form) or filter wedges for achieving equality of intensity in the split field of views of the instrument.

6.2.2 Visual Illuminometers

The photometers discussed above are mainly for measuring the luminous intensity or flux, or comparing two sources of luminous energy. Illuminometers, on the other hand, are designed for measuring the luminous energy reaching a surface such as a page of a book, a work table, the surface of a street, or other object to be viewed by the eye. Since plant responses, such as photosynthesis and growth factors, are controlled mainly by radiation in the visible region, plant scientists have long been interested in measurements of natural illumination. Most modern instruments for the measurement of illumination are based on photoelectric devices, but visual illuminometers have a historical interest and are still used in some applications. For instance, the Bylov model of the basic Weber photometer is used for measuring natural illumination in the Soviet Union (Kondratyev, 1969).

Of the several visual illuminometers which have been developed, the best known in the West is perhaps the Macbeth illuminometer.

Macbeth Illuminometer   This instrument uses the basic Lummer–Brodhun head (see Fig. 6.4 on p. 169) to compare the light reflected from a matte-white test plate positioned at the point of measurement with that from a translucent opal glass comparison plate illuminated with an incandescent lamp. The distance between lamp and comparison plate is varied by means of a rack-and-pinion for equalizing the intensities in the two parts of the viewed field. A scale, which has been calibrated in terms of illumination, is attached to the shaft which controls the lamp position. The power supply with rheostat and milliammeter for energizing the lamp is normally carried in a separate control case. Neutral density filters may be inserted in either of the light paths in order to extend the range of the instrument from its normal scale range of 1–25 foot-candles to a range of 0.01–2500 foot-candles.* A special color filter is available for equalizing the color or normal daylight illumination to that of the reference lamp.

The Macbeth illuminometer may be used to measure the illumination on a surface, the brightness of an extended source, or the luminous intensity of a small area source. It is portable and simple to operate, and is often used for measurements of both natural and artificial illumination.

The accuracy attainable with visual photometers or illuminometers

---

* Information kindly supplied by Leeds and Northrup, San Mateo, Calif., 1973.

is nearly as high as that of most radiation measurements. Under ideal laboratory conditions with experienced observors, accuracies of the order of 1–2% are possible with the Lummer–Brodhun photometer or the Macbeth illuminometer. With simpler devices and under field conditions, the attainable accuracy is degraded by a factor of 2 or more.

6.2.3   Photoelectric Illuminometers

As mentioned above, the versatility, adaptability, and convenience of modern photoelectric devices have caused them to largely supplant visual measurements of illumination. The devices fall into the general categories of photoemissive, photoconductive, and photovoltaic cells. The first of these, including the photomultiplier, is useful for measurements at very low light levels. Photoconductive cells are also relatively sensitive and are useful for extending measurements into the infrared region of the spectrum. Neither of these is required, however, for measuring daylight illumination for meteorological purposes. In addition, both photoemissive and photoconductive cells require an external power supply for supplying a bias voltage.

By far the most convenient of the devices for measuring illumination is the barrier-layer photovoltaic cell, and numerous types of light-sensitive devices (photographic light meters, light-actuated switches, etc.) ultilize such cells in their operation. They are the basis of most modern illumino-meters, and portable illuminometers (often termed foot-candle meters) are available from many different manufacturers.

The barrier-layer cell is an extremely simple device. It consists of a sandwich configuration of a metal base (iron or copper) covered with a layer of a semiconductor (cuprous oxide or selenium), on top of which is sputtered a thin film of another metal (gold, silver, etc.). Photons pene-trating the top metal film cause the migration of electrons from the semi-conductor to the metal film, which, when collected at appropriate contacts, produce a measurable current through an attached microammeter.

According to the criteria of spectral sensitivity, temperature depen-dence, linearity of response, and adherence to the cosine law, the selenium cell with a correcting filter and appropriately mounted diffusing plate performs very well for measurements of illumination. The spectral sensi-tivities of a selenium cell without a filter and a selenium cell with a Weston Viscor filter are compared with that of the standard light-adapted human eye in Fig. 6.5. The correspondence between the filtered selenium cell and the eye is certainly adequate for practical purposes.

The effect of temperature on the sensitivity of selenium cells has

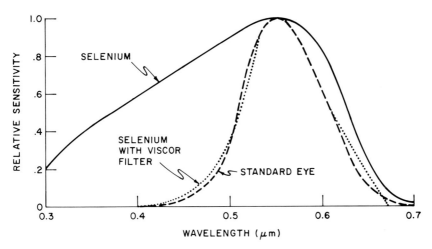

Fig. 6.5   Spectral sensitivities of a selenium barrier-layer photocell and a selenium cell with Weston Viscor filter compared with that of the standard light-adapted human eye.

been studied by Foster (1951), who found that cells of manufacture before about 1950 had very little temperature dependence for temperatures below 32° to 34°C, but it increased rapidly at higher temperatures and reached -20% at about 50°C. Later cells are improved sufficiently, however, that if the cell is calibrated at an intermediate temperature, it is possible to achieve only a $\pm$ 2–3% temperature error over the entire temperature range of $-56°$ to $+50°C$. Foster found that the stability of the cell was greatly improved by enclosing it in a hermetically sealed case, into which was pumped a dry inert gas.

The response of the selenium cell is linear with changing light intensities if the light intensity and external resistance in series with the cell are both relatively low. For this reason, most illuminometers are fitted with screens for assuring sufficiently low light levels on the cell. An external resistance of not greater than 200 Ω is recommended, in which case a current of up to 200 μA can be drawn from the cell and still maintain a linear response. The response is also somewhat nonlinear and sensitivity problems are encountered at very low light levels ($<$ 0.1 foot-candles). Various methods of overcoming these problems have been discussed by Horton (1969), who shows that the performance at low light levels is greatly improved by the use of feedback circuits or solid-state amplifiers in the output of the cell.

As with other radiation instruments, the deviation of the response from the ideal cosine law with varying angles of incidence is a problem with selenium photocells. For uncorrected cells, this cosine effect may cause errors of 30% or more at angles of incidence beyond 60°. Even greater errors result if the cell is mounted in a holder with a rim which casts a shadow on the sensitive surface. A method of minimizing the cosine error for obliquely incident light by the use of a diffusing disk projecting above the mount was developed by Pleijel and Longmore (1952). The method is used in many selenium photocell illuminometers such as that of the Eppley Laboratories. The diffusing disk placed over the sensor is made to project above the mounting ring such that part of the light at large angles of incidence enters the exposed edge of the disk and is scattered down to the sensing surface. By a combination of contouring the exposed edge and adjusting its height, the cosine error can be reduced to 2% or less for all angles of incidence.

One troublesome feature of selenium barrier-layer photocells in continuous operation for measuring illumination is a gradual decrease of sensitivity due to fatigue of the cell. The fatigue can be minimized by restricting the light intensity on the cell to low levels, but it is not eliminated. Because of the fatigue effect, frequent calibrations of selenium photocell illuminometers are necessary.

6.3 MEASUREMENTS OF NATURAL ILLUMINATION

Daylight illumination is not one of the parameters measured at ordinary meteorological stations, and it is not included in the climatological data summary of the United States. Illumination is measured on a routine basis, however, at a number of observatories (e.g., Kew, University of Bergen, University of Moscow, etc.) and instructions were given for its measurement at various locations during the International Geophysical Year. An analysis of a series of measurements of illumination in South Africa has been provided by Drummond (1956, 1958). Available information indicates that illuminometers based on the selenium barrier-layer photocell are used in all cases of continuous recording.

As would be expected, the illumination on a horizontal surface due to solar radiation under clear skies is a relatively strong function of Sun elevation. In fact, observations at the Geophysical Institute of the University of Bergen (Schieldrup–Paulsen, 1970) indicate a linear dependence of illumination on Sun elevation up to at least elevations of 50°. Data given by Drummond (1958) confirm the essentially linear dependence for Sun elevations of 50° or less, but the dependence is progressively weaker from 50°

to 90°. The diurnal variation of natural illumination on a horizontal surface for cloudless skies at Pretoria, South Africa, as measured by Drummond (1958), is shown in Table 6.1.

Midday values range from about 64 to 119 klx-hr between winter and summer. The equivalent range for the Kew Observatory located 16 km west of London is about 20 to 100 klx-hr. Extreme values of 120 and 130 klx-hr have been observed at Kew and Pretoria, respectively, under conditions of clouds near but not obscuring the Sun. The midday contribution due to diffuse skylight alone was found to be almost constant at 10–11 klx-hr throughout the year at Pretoria.

Because of the relative abundance of measurements of the total flux of solar radiation on a horizontal surface and the paucity of illumination measurements, it would be useful for practical purposes to interpret the radiation measurements in terms of their equivalent illumination. This conversion could be performed directly by the use of Eq. (6.7) if the spectral distribution of the incident energy were known. Generally, however, such detailed information is not available, since routine solar flux measurements are made with broad-band pyranometers.

TABLE 6.1

The Diurnal Variation of Natural Illumination on a Horizontal Surface for Cloudless Skies at Pretoria, South Africa[a]

Hour (local apparent time)	Month			
	March	June	Sept.	Dec.
0600–0700	8.7	0.5	7.4	26.6
0700–0800	32.8	12.0	27.2	51.1
0800–0900	59.1	29.8	49.3	73.6
0900–1000	80.0	44.3	69.2	94.1
1000–1100	96.2	56.0	84.1	110.3
1100–1200	105.1	64.0	90.9	118.8
1200–1300	103.8	64.2	89.5	118.0
1300–1400	93.7	58.2	81.0	109.4
1400–1500	76.2	45.3	65.9	92.7
1500–1600	54.5	29.2	45.8	71.6
1600–1700	30.3	10.5	24.2	47.7
1700–1800	8.3	0.5	5.4	24.8
Total	748.7	414.5	639.9	938.7

[a] As measured by Drummond (1958). The units are kilolux-hours. Pretoria is at 25.8° S latitude at an altitude of 1370 m.

A second difficulty arises because of large variations of atmospheric absorption in the near-infrared bands of water vapor and carbon dioxide. These variations strongly affect the total energy flux reaching the surface, but they have little or no effect on the flux of visible radiation. A. K. Ångström and Drummond (1962) estimate the conversion factor for direct solar radiation to vary by $\pm$ 10% owing to water vapor absorption alone. No equivalent estimate is available for the total global radiation, but it is likely that the increased atmospheric pathlength for scattered light would cause larger variations for the skylight component than that for the direct component. Thus it appears that only relatively gross determinations of the illumination can be made by the use of data from pyranometers operating in the normal manner without spectral filters.

The accuracy of the determination of illumination is much improved, however, for radiometric measurements with broad-band filters such as the three standard Schott OGI, RG2, and RG8 cutoff filters. Drummond and A. K. Ångström (1971) have shown that the most accurate determinations are obtained by the use of the RG8 filter, which gives radiation fluxes integrated over the spectral interval of $\lambda < 0.70$ μm. This is a logical result, as all of the visible radiation is included but the water vapor effects in the near infrared are avoided. According to those authors, the luminous flux L in units of kilolux due to global radiation flux F at $\lambda < 0.70$ μm expressed in cal cm^{-2} min^{-1} is given by the relation

$$L = 150F(1 + 0.102m) \qquad (6.9)$$

where m is absolute air mass (i.e., relative air mass reduced to sea level). A comparison of results obtained by this relation with those of 67 series of measurements under various atmospheric conditions gave an average error of only 1% in the determination of L. Drummond and Ångström (1971) concluded that the results confirmed the wide generality of Eq. (6.9) for use on a practical basis.

In spite of the astounding accuracy indicated in this empirical relationship, the utility of the method for practical application would seem to be limited. Most radiometric stations are not equipped with a filtered pyranometer, and the relation is applicable only to radiation confined to the region at $\lambda < 0.70$ μm. Furthermore, since the RG8 filter transmits radiation at $\lambda > 0.70$ μm, the measurement of that at $\lambda < 0.70$ μm requires the use of two pyranometers, one with the RG8 filter and one without a filter. Only the larger observatories would normally be provided with such a complete complement of instruments, and it is likely that those observatories could also include an illuminometer if information on the illumination were desired.

REFERENCES

Ångström, A. K., and Drummond, A. J. (1962). Fundamental principles and methods for the calibration of radiometers for photometric use. *Appl. Opt.* **1**, 455–464.

Condon, E. U., and Odishaw, H. (1958). "Handbook of Physics," 2nd ed. McGraw–Hill, New York.

Drummond, A. J. (1956). Notes on the measurement of natural illumination; I. Some characteristics of illumination recorders. *Arch. Meteorol. Geophys. Bioklimatol.* **7**, 437–465.

Drummond, A. J. (1958). Notes on the measurement of natural illumination; II. Daylight and skylight at Pretoria; the luminous efficiency of daylight. *Arch. Meteorol. Geophys. Bioklimatol.* **9**, 149–163.

Drummond, A. J., and Ångström, A. K. (1971). Derivation of the photometric flux of daylight from filtered measurements of global (Sun and sky) radiant energy. *Appl. Opt.* **10**, 2024–2030.

Foster, N. B. (1951). A recording daylight illuminometer. *Illum. Eng. New York* **46**, 59–62.

Horton, G. A. (1969). Electric instrumentation in light measurement. *Illum. Eng.* **64**, 701–707.

Kondratyev, K. Ya. (1969). "Radiation in the Atmosphere." Academic Press, New York.

Pleijel, G., and Longmore, J. (1952). A method of correcting the cosine error of selenium rectifier photocells. *J. Sci. Instrm.* **29**, 137–138.

Schieldrup-Paulsen, H. (1970). "Radiation Observations in Bergen, Norway." Radiation Yearbook No. 6. Geophysical Institute, University of Bergen, Bergen, Norway.

Viezee, W. (1960). Survey of radiometric quantities and units. Res. Memo. RM–2492, The Rand Corporation, Santa Monica, California.

Walsh, J. W. T. (1958). "Photometry." Constable Press, London.

CHAPTER SEVEN

Polarization of Light in the Atmosphere

7.1 INTRODUCTION

A general discussion of the polarization of light is outside the scope of this book, but the subject is covered thoroughly in many standard works on radiation and optics (cf. Wood, 1934; Born and Wolf, 1959; Stone, 1963). The polarization of skylight, however, has received much less attention than has the more general subject of the polarization of light, and it is still an area of very active research. More progress has been made in both theory and measurements of the polarization of skylight during the last two or three decades than was made in the previous 80 years since Lord Rayleigh's famous explanation of molecular atmospheric scattering in 1871. Of the many reasons for the renewed interest in skylight polarization, probably the most important are, first, the development of powerful new theoretical techniques in radiative transfer (Ambartsumian, 1943, 1944; Chandrasekhar, 1950; Sobolev, 1949, 1950), and second, the advance of technology by which polarization can be easily measured by modern electronic instrumentation. Also here, as in almost every field, the development of electronic computers opened up new possibilities of numerical integration, which have been exploited to good advantage in solving the equation of radiative transfer in realistic models of the atmosphere. The main results of these investigations are summarized in the sections which follow.

The history of investigations of the polarization of skylight up to 1950 has been well summarized by Sekera (1950), so only the highlights will be mentioned here. The fact that the light which reaches us from the sunlit sky is partially polarized was first discovered by the French physicist D. F. J. Arago (1786–1853) in 1809. Arago then went on to establish the position of the maximum of the polarization field as being in the direction at approximately 90° from the direction of the Sun and to discover the existence of a neutral point (point of zero polarization) at a position in the sky at about 25° above the antisolar direction (direction exactly opposite that of the Sun), which has been named for him. The other two neutral points which normally occur in the sunlit sky were discovered by Jacques Babinet (1794–1872) and Sir David Brewster (1781–1868) in 1840 and 1842, respectively.

Attempts to explain the observed field of skylight polarization in a clear atmosphere yielded the theory of molecular scattering by Lord Rayleigh (Strutt, 1871), by which several, but not all, of the observed features can be explained. Rayleigh's elementary theory explained neither the existence of the neutral points nor the lack of complete polarization at the maximum for the completely clear atmosphere.

Little was known in the 1870's about the scattering of sunlight by atmospheric aerosols. Observations after the eruption of Krakatao in 1883 showed, however, that the magnitude of skylight polarization is generally decreased and a significant shift of the positions of the neutral points is produced by volcanic ash in the upper atmosphere. The same features were observed after the eruption of Mt. Katmai in 1912. In both cases attempts at theoretical explanation of skylight observations were made, the main investigators being Soret (1888) after Krakatao, and King (1913) after Mt. Katmai.

Although the gap between Rayleigh's theory of molecular scattering and the well-known principles of geometrical optics was successfully bridged by the German physicist Gustav Mie in 1908, the potential of the theory for explaining polarizing effects of atmospheric aerosols came to fruition only after the advent of the electronic computer. In fact, because of complexities of the theory, lack of knowledge of characteristics of the aerosols, and limitations of present day computers, we have still not been able to apply the full power of the Mie theory to problems of multiple scattering of light in the atmosphere, although the primary scattering problem is now well in hand.

Before about the time of World War II, measurements of skylight polarization were mainly by visual means. Probably the most accurate of the visual polarimeters and one which enjoyed considerable popularity is the Cornu–Martens polarimeter (Martens, 1900). The Savart polariscope was frequently used for determining the positions of the neutral points. Both of these instruments are briefly described in Section 7.5. The development of sensitive photodetectors, particularly the multiplying phototubes, and sophisticated electronics opened up a new era in skylight measurements. The new techniques have been employed by several investigators since World War II, including Sekera *et al.* (1955), Sekera (1956), Bullrich *et al.* (1966), Gehrels and Teska (1963), Coulson and Walraven (1972), and others. The details of modern electronic polarimeters are discussed in Section 7.5.

7.2 POLARIZATION OF RADIATION

The degree of polarization of a stream of radiation is defined in terms of the four Stokes parameters by the relation

$$P = (Q^2 + U^2 + V^2)^{1/2}/I \qquad (7.1)$$

where the total intensity I is given as the sum of the orthogonal intensity components I_e and I_r:

$$I = I_e + I_r \qquad (7.2)$$

The parameters Q, U, and V are given as

$$Q = I_e - I_r \qquad U = Q \tan 2\chi \qquad V = I \sin 2\beta \qquad (7.3)$$

where χ is the angle between the plane of polarization and the direction of the e coordinate. The parameter β is

$$\beta = \arctan(b/a) \qquad (7.4)$$

a and b being proportional, respectively, to the lengths of the major and minor axes of the ellipse described by the end of the vibrating electric vector. If the four quantities I, Q, U, and V are known, the shape and orientation of the ellipse can be determined, thereby completely defining the state of polarization. For unpolarized (neutral) radiation, $Q = U = V = 0$; for only linear polarization, $V = 0$; for only circular polarization $Q = U = 0$ but $V \neq 0$; for completely polarized radiation $I^2 = Q^2 + U^2 + V^2$. The most general type of radiation stream is composed of a combination of unpolarized radiation and elliptically polarized radiation. For this case, all of the Stokes parameters are nonzero, and the relation

$$I^2 > Q^2 + U^2 + V^2 \qquad (7.5)$$

is valid. Obviously we have the constraint

$$0 \lesssim P \lesssim 1 \qquad (7.6)$$

It is observed that the circularly or elliptically polarized component of the light in the atmosphere is very small compared to the plane polarized and neutral components, in which case we can, for practical purposes, set $V = 0$ for atmospheric scattering problems. Thus the physically important characteristics of the field of skylight are given completely in terms of the intensity I, the degree of (plane) polarization P, and the angle χ between the plane of polarization and the vertical direction.

7.3   THEORETICAL DETERMINATIONS OF POLARIZATION

For application of scattering theory to the case of the sunlit sky, we assume the incident solar radiation at the top of the atmosphere to be parallel, with a net flux of energy in unit time and unit frequency interval across a unit surface oriented normal to the incident beam to be $\pi \mathbf{F}_0$, where $\mathbf{F}_0$ is a one-column matrix of the Stokes parameters written as

$$\mathbf{F}_0 = \begin{bmatrix} F_{0\,\|} \\ F_{0\perp} \\ U_0 \\ V_0 \end{bmatrix} \tag{7.7}$$

Measurements show the extra-atmosphere incident radiation to be unpolarized, for which case we have the relations

$$F_{0\,\|} = F_{0\perp} = \tfrac{1}{2}F_0$$
$$U_0 = V_0 = 0 \tag{7.8}$$

A portion of this incident radiation which is scattered by gaseous molecules and other particles in the atmosphere eventually shows up as diffuse light from the sunlit sky. The vector intensity $\mathbf{I}$ of the radiation scattered by a spherical particle is given in its most general form by the relation

$$\mathbf{I} = \begin{bmatrix} I_{\|} \\ I_{\perp} \\ U \\ V \end{bmatrix} = \mathbf{P}F_0 \tag{7.9}$$

where $\mathbf{P}$ is a 16-element scattering matrix with elements P_{ij}, $i, j = 1, 2, 3, 4$. It can be shown (van de Hulst, 1957) that for spherical particles all of the elements of the scattering matrix except six are zero, thereby yielding the matrix

$$\mathbf{P} = \begin{bmatrix} P_{11} & 0 & 0 & 0 \\ 0 & P_{22} & 0 & 0 \\ 0 & 0 & P_{33} & P_{34} \\ 0 & 0 & P_{43} & P_{44} \end{bmatrix} \qquad (7.10)$$

and, further, that $P_{33} = P_{44}$ and $P_{34} = -P_{43}$. This reduces the number of independent elements to four. By means of the Mie theory of scattering, we have the relations, where * denotes complex conjugate,

$$P_{11} = S_1 S_1{}^*$$

$$P_{22} = S_2 S_2{}^*$$

$$P_{33} = P_{44} = \text{Re}\{S_1 S_2{}^*\} \qquad (7.11)$$

$$P_{34} = P_{43} = \text{Im}\{S_1 S_2{}^*\}$$

Here S_{ij} are the amplitude functions of the Mie formulation. They will be mentioned further in connection with aerosol effects in the atmosphere.

For Rayleigh scattering, in which case the particles are small with respect to the wavelength, the vector intensity of the scattered radiation is given, with respect to scattering angle Θ, by the simplified relation (Sekera, 1957)

$$I = \left(\frac{2\pi}{\lambda}\right)^4 \Delta^2 \begin{bmatrix} \cos^2\Theta & 0 & 0 & 0 \\ 0 & 1 & 0 & 0 \\ 0 & 0 & \cos\Theta & 0 \\ 0 & 0 & 0 & \cos\Theta \end{bmatrix} F_0 \qquad (7.12)$$

Δ is the depolarization factor, to be discussed later.

The geometry of the scattering problem can be seen from Fig. 7.1a. The scattering particle is assumed to be at 0, the unit vectors parallel and normal to the scattering plane (plane defined by directions of incident and scattered light) are indicated by $\parallel$ and $\perp$, respectively, and the scattering angle is Θ.

A clear physical picture of the distribution of intensity around a scat-

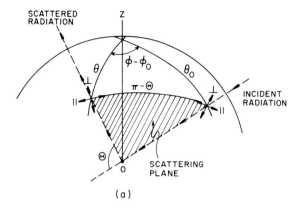

(a)

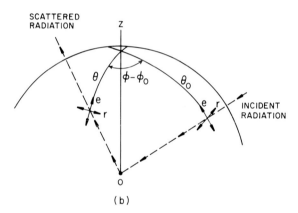

(b)

Fig. 7.1 (a) Geometry of scattering for the case of a coordinate system referred to the plane of scattering. Symbols $\parallel$ and $\perp$ indicate directions parallel and normal, respectively, to the scattering plane (shaded). Θ is scattering angle. (b) Geometry of scattering for a coordinate system referred to the meridional planes at the azimuths of incident and scattered radiation.

tering particle is perhaps best seen in terms of a scalar phase function $P(\Theta)$ instead of the complete phase matrix. For the case of Rayleigh scattering, the distribution of (scalar) intensity is given by the phase function

$$P(\Theta) = \tfrac{3}{4}(1 + \cos^2 \Theta) \tag{7.13}$$

The factor $\tfrac{3}{4}$ is necessary to normalize to unity the phase function integrated over the entire solid angle. The intensity distribution surrounding a molecule scattering according to Eq. (7.13), represented by the solid curve

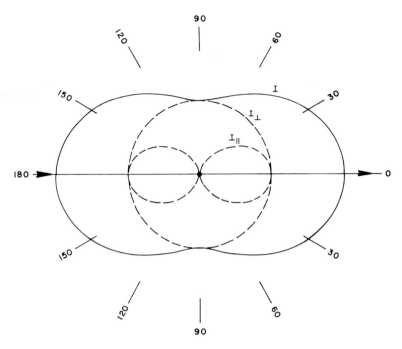

Fig. 7.2 Scattering pattern of a Rayleigh particle for unpolarized incident radiation. $I_{||}$ and $I_\perp$ are the components polarized parallel and normal, respectively, to the scattering plane. The pattern has axial symmetry around the direction of propagation of the incident radiation.

of Fig. 7.2, shows the following relative magnitudes, where subscripts indicate values of Θ in degrees: $I_{180} = I_0$; $I_{90} = \frac{1}{2}I_0$; for $0° < \Theta < 90°$ and $90° < \Theta < 180°$, $\frac{1}{2}I_0 < I_\Theta < I_0$. The distribution has rotational symmetry around the $0°$–$180°$ direction.

If the vibrations of the electric vector of the scattered radiation are resolved into the orthogonal direction parallel and perpendicular to the scattering plane, shown as shaded in Fig. 7.1a, the Stokes parameters of the scattered radiation are $(I_{||}, I_\perp, U, O)$ for unpolarized incident radiation. $V = 0$ because the scattering of unpolarized incident light does not introduce elliptic polarization (van de Hulst, 1957). The total intensity is $I = I_{||} + I_\perp$, and the components have the characteristics $I_\perp = \text{const}$ and $I_{||} \sim \cos^2 \Theta$, shown by the dashed curves of Fig. 7.2.

The degree of polarization of the scattered light is

$$P = (I_\perp - I_{||})/(I_\perp + I_{||}) \tag{7.14}$$

and its variation with Θ is given by

$$P(\Theta) = (1 - \cos^2 \Theta)/(1 + \cos^2 \Theta) = \sin^2 \Theta/(1 + \cos^2 \Theta) \qquad (7.15)$$

As shown in Fig. 7.3, light scattered in the forward and backward direction is completely unpolarized ($P = 0$), light scattered at $\Theta = 90°$ is completely polarized ($P = 1$), and light scattered at all other angles is partially polarized ($0 < P < 1$).

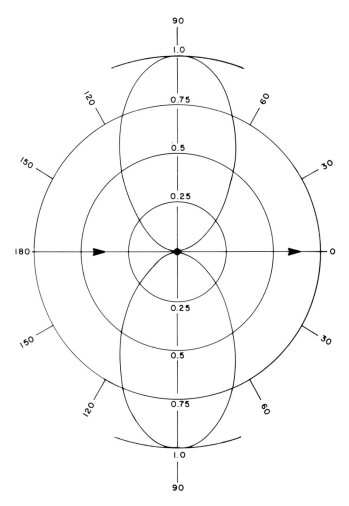

Fig. 7.3 Angular distribution of the degree of polarization of radiation scattered by a Rayleigh particle for the case of unpolarized incident radiation. The pattern has axial symmetry around the direction of propagation of the incident radiation.

In applying these relations to an actual planetary atmosphere, such as that of the Earth, it is more convenient to use the coordinate system shown in Fig. 7.1b, in which the relevant angles are zenith angles θ and θ_0 and azimuth angles ϕ and ϕ_0 for scattered and incident radiation, respectively. For this system the Stokes parameters are (I_e, I_r, U, O), I_e and I_r being the components in and normal, respectively, to the vertical plane at the appropriate azimuth. The relations given above as Eqs. (7.1)–(7.5) apply to this more convenient coordinate system.

One of the assumptions in Rayleigh's analysis is that the scattering molecules are isotropic in character. In its most general form, a particle may be asymmetric along all three of its axes, in which case its polarizability has the form of a tensor, the three main components of which can be obtained from an analysis of the dipole moments of the particles (see van de Hulst, 1957). By taking averages for all possible orientations among many particles one can find two quantities A and B by which the scattering matrix can be expressed. In the interest of brevity, this analysis is omitted here; it is well developed by van de Hulst. On this basis, the depolarization factor Δ is given by the relation

$$\Delta = (2A - 2B)/(4A + B) = I_e(\Theta = 90°)/I_r(\Theta = 90°) \qquad (7.16)$$

for the case of unpolarized incident light. For this restricted but very useful case the phase function of Eq. (7.13) is modified to

$$P(\Theta) = [3/2(2 + \Delta)][1 + \Delta + (1 + \Delta)\cos^2\Theta] \qquad (7.17)$$

Although the value of Δ is readily determined by laboratory measurements, there is considerable disparity among values given by different authors. A commonly accepted value is 0.031 for dry atmospheric air, in which case the degree of polarization for right-angle scattering is reduced from $P(90°) = 1.0$ to $P(90°) = 0.969/1.031 = 0.940$, a fact which is partly responsible for the considerably less than complete polarization observed in the atmosphere.

Another assumption on which Rayleigh's theory was based is that of primary (single) scattering, which is applicable to the case of one or a small group of molecules. The Earth's atmosphere, however, contains such large numbers of molecules and aerosol particles that a photon may well be scattered two, three, or more times during its path in the atmosphere. Dave (1964) has shown that in order to get good agreement of theory with observations even for the pure molecular (Rayleigh) atmosphere of the Earth, it is necessary to take into account two orders of scattering for wavelength $\lambda = 0.644$ μm, three for $\lambda = 0.546$ μm, and up to eight orders of scattering for $\lambda = 0.312$ μm. Although the problem of multiple scattering among aerosol particles and between aerosol particles and molecules of the air has not yet been completely solved, it is known that multiple scattering in such a turbid atmosphere is extremely important. Strong evidence for this will be given below for the case of a polluted atmosphere. Similarly, in a dense atmosphere, such as that of Venus, and in water clouds in the Earth's atmosphere a representative photon may undergo tens or even hundreds of scattering events during its residence in the atmosphere, while for the tenuous atmosphere of Mars accounting for second-order scattering is sufficient, even for ultraviolet radiation.

The effects of multiple scattering on the intensity of skylight were discussed in Chapter 4; here we confine ouselves to the polarization field. The lack of complete polarization for right-angle scattering could be qualitatively explained by molecular anisotropy, but the measurements of Jensen (1942), Dorno (1919), and others showed still lower polarization values than could be explained by molecular anisotropy, even for the clearest atmosphere. Furthermore, Rayleigh's theory for primary scattering predicted neutral points (points of zero polarization) in only the exactly forward and backward directions, but neutral points had already been observed above and below the Sun and above the anti-Sun by Babinet, Brewster, and Arago several decades before the time of Rayleigh. Soret (1888) had tried, with qualitative success, to explain the neutral points by higher-order scattering, but the convincing proof had to wait still another 60 years for the work of Chandrasekhar (1950) and Chandrasekhar and Elbert (1951). By taking account of all orders of scattering in a Rayleigh atmosphere, the existence and observed behavior of the neutral points are explained, as is the further decrease of the polarization maximum and its dispersion (variation with wavelength).

The parameters which have been found to best characterize the polarization field of skylight are the magnitude of the maximum polarization, the position of the maximum polarization with respect to the position of the Sun, and the positions of the three neutral points.

7.3.1   Polarization in a Rayleigh Atmosphere

The polarization field for the case of a Rayleigh atmosphere has been studied extensively during the last two decades by many authors, including Sekera (1950, 1956), Coulson (1959, 1969a,b), de Bary (1964), de Bary and Bullrich (1964), Dave (1964), Dave and Furukawa (1966), Fraser (1964), Gehrels (1962), and others. The data for the figures below were taken from the tables of Coulson, Dave, and Sekera (1960).

The degree of polarization in the principal plane (vertical plane through the Sun) for a Rayleigh atmosphere is shown as a function of zenith angle for four different wavelengths (at height of sea level) in Fig. 7.4. The de-

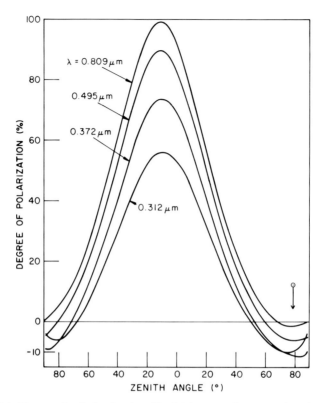

Fig. 7.4 Degree of polarization in a Rayleigh atmosphere as a function of zenith angle in the plane of the Sun's vertical at four different wavelengths for a solar zenith angle of 78.5° $(A = 0)$.

polarization factor [Eq. (7.10)] has been neglected in these and subsequent curves. The effects of multiple scattering are clearly evident in the decrease of positive polarization and increase of negative polarization with decreasing wavelength. The physical reason for multiple scattering changing the degree of polarization is that in a statistical sense the photons which are incident on a given scatterer, after having been scattered one or more times, arrive from all possible directions, so those scattered into a given direction by the scatterer encompass all possible scattering angles. This means that all magnitudes of polarization $(0 \lesssim P \lesssim 1)$ are produced in the composite stream. Since, statistically, $P < 1$ in the multiple-scattered component, multiple scattering can only reduce the degree of polarization in the direction at right angles to the source, for which $P = 1$ for primary scattering. Conversely, in those directions in which primary scattering gives $P = 0$, multiple scattering introduces radiation characterized by the intensity component $I_{\parallel}$ parallel to the scattering plane as being greater than the intensity $I_{\perp}$ normal to the scattering plane. By convention this is called negative polarization. Thus the composite stream at scattering angles $\theta = 0°$ or $180°$ is changed from neutral ($P = 0$) for primary scattering to partially polarized for the composite. The neutral points of the field occur in those directions at which the positive polarization resulting from primary scattering is just balanced by the negative polarization of the multiple-scattered component, each being weighted, of course, by the relative intensity of its contribution to the total. Symbolically, we can write for the neutral points

$$(PI)_{\text{primary}} + (PI)_{\text{multiple}} = 0 \qquad (7.18)$$

On this basis, the increasing shift of the neutral points away from $\theta = 0$ with decreasing wavelength, as shown in Fig. 7.4, can be understood to result from a relative increase of I_{multiple} and relative decrease of I_{primary} with decreasing wavelength, and there is no requirement for a wavelength dependence of P for the individual streams.

The polarization pattern moves across the sky as the Sun changes position, as shown in Fig. 7.5. This is a manifestation of the fact that the scattering angle θ for primary scattering is the dominant parameter for this case of a Rayleigh atmosphere of moderate optical thickness. A number of other changes of the polarization field with position of the Sun are evident in the diagram. First, the degree of polarization at the maximum, P_{max}, reaches higher values at very high solar zenith angles θ_0 than at moderate values of θ_0. This effect has been studied by Coulson (1952), who found that P_{max} for this case shows a broad minimum at $\theta_0 = 50°$ to $60°$, with its value increasing by about 1% at $\theta_0 > 80°$ and by as much as 4 to 5% at small values of θ_0. This variation must be due to a complex of the polarizing effects of primary and multiple scattering in combination with the relative incident intensity and transmission of the scattered light through varying optical paths as the viewed direction moves between the zenith and horizontal directions.

It is known that radiation which is reflected from the surface and scattered back down toward the surface has a significant effect on the polarization field of skylight. A theoretical method for accounting for this component of skylight was developed by Chandrasekhar (1950), and many authors, including Chandrasekhar and Elbert (1954), Coulson (1968), Deirmendjian and Sekera (1954), and others have studied the problem. The computations are particularly simple for the case of a Lambert surface (reflected radiation unpolarized and isotropic in the outward hemisphere), although Coulson (1968) has discussed the effects of reflection from a real

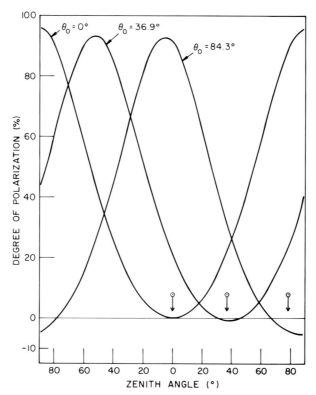

Fig. 7.5 Degree of polarization in a Rayleigh atmosphere as a function of zenith angle in the plane of the Sun's vertical at a wavelength of 0.546 μm for three different zenith angles of the Sun ($A = 0$).

desert surface, and the case of reflection from a Fresnel (mirror-type) surface has been discussed by Fraser (1964).

Curves of skylight polarization in the principal plane for a Rayleigh atmosphere overlying a Lambert surface are shown in Fig. 7.6. Results for total reflectances (albedos) typical of a completely black surface ($A = 0$), a desert surface ($A = 0.25$), and a new snow surface ($A = 0.80$) are given. An interpretation of the effects introduced by surface reflection can be obtained by considering the stream of skylight radiation to be composed of that which would exist in the absence of a ground or for a black ground ($A = 0$), and that which results from surface reflection. The degree of polarization P of the composite is given for the principal plane by the relation

$$P = [(I_e - I_r) + (I_e^* - I_r^*)]/(I_e + I_r + I_e^* + I_r^*) \qquad (7.19)$$

where the asterisks indicate the contribution by surface reflection. Since the outward directed light at the surface is assumed unpolarized and is of a diffuse character, scattering by the atmosphere introduces little polarization into this component. Thus to a first approximation $I_e{}^* - I_r{}^* = 0$, in which case reflection does not affect the numerator of Eq. (7.20). But $I_e{}^* + I_r{}^* \neq 0$ in general, and the denominator of Eq. (7.20) may be significantly increased by surface reflection, with the result shown in Fig. 7.6. Since, by definition, the addition of the unpolarized light contributed by reflection in the direction of the neutral points of the original field would produce no change of polarization, the positions of the neutral points would be very insensitive to Lambert surface reflection. This expected insensitivity is seen in the curves.

Coulson (1968) has shown that for a Rayleigh atmosphere overlying a desert surface, for which the measured reflection characteristics are taken

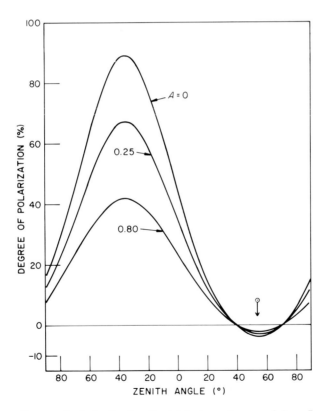

Fig. 7.6 Degree of polarization in a Rayleigh atmosphere overlying a Lambert-type surface for three different values of surface albedo ($\lambda = 0.436 \ \mu m$, $\theta_0 = 53.1°$).

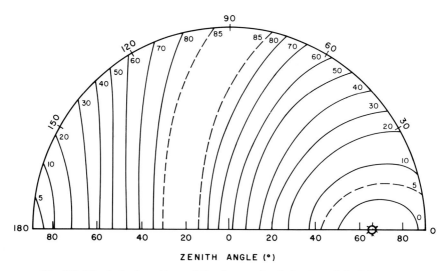

Fig. 7.7 Hemispheric pattern of the degree of polarization of skylight at a wavelength of 0.495 μm in a Rayleigh atmosphere ($\theta_0 = 66.4°$; $A = 0$). Only one-half of the hemisphere is shown; the other half is a mirror image of this one.

into account, the degree of polarization of the backscattered radiation is not greater than 1–3%, for which case the Lambert surface assumption is a good approximation for the polarization of skylight. It is not a good approximation, however, for the light emerging to space from the top of the atmosphere, as surface effects are much more important for the outward radiation than for skylight.

The pattern of polarization over the entire hemisphere of the sky is shown in Fig. 7.7 for a Rayleigh atmosphere in the absence of surface reflection. The position of the polarization maximum is maintained at a total scattering angle of approximately 90° from the Sun, and its magnitude is practically independent of position in the sky. There is a considerable region of negative polarization surrounding the position of the Sun, and the positions of the Babinet and Brewster points, above and below the Sun, respectively, are well defined in the pattern. At other solar positions (not shown) the polarization field is shifted corresponding to the shift of the Sun, but it is qualitatively similar to that of Fig. 7.7. Likewise, the general hemispheric pattern is largely independent of wavelength, although the degree of polarization is higher at longer wavelengths and lower at shorter wavelengths than that shown in the diagram.

7.3.2 Polarization Effects of Atmospheric Aerosols

The real atmosphere always contains greater or lesser numbers of dust, haze, and other types of particles, in addition to the molecular size particles of the atmospheric gases. These aerosol particles affect the solar radiation regime significantly, even over the far reaches of the open ocean, so their effects must be included in any realistic assessment of the radiative transfer problem. The scattering of radiation by spherical particles of radius comparable to or larger than the wavelength of the radiation was solved by Gustav Mie in 1908. The theory is not simple, and any attempt at a comprehensive discussion of it is beyond the scope of this book. It has been discussed in detail by a number of authors, e.g., van de Hulst (1957), and Deirmendjian (1969). For convenience, however, the basic expressions will be written below.

The fundamental relations required are a pair of coefficients, the Mie coefficients a_n and b_n, which are independent of the scattering angle θ, and a pair of functions $\pi_n(\cos\theta)$ and $\tau_n(\cos\theta)$, involving Legendre polynomials and their derivatives, which are dependent on only the scattering angle. The Mie coefficients are the following:

$$a_n = \frac{\psi_n'(y)\psi_n(x) - m\psi_n(y)\psi_n'(x)}{\psi_n'(y)\xi_n(x) - m\psi_n(y)\xi_n'(x)}$$

$$b_n = \frac{m\psi_n'(y)\psi_n(x) - \psi_n(y)\psi_n'(x)}{m\psi_n'(y)\xi_n(x) - \psi_n(y)\xi_n'(x)}$$

(7.20)

Here x is given by $x = 2\pi r/\lambda$, where r is particle radius and λ is wavelength of the radiation; m is relative index of refraction of the particles; and $y = mx$. The functions ψ_n and ξ_n are Riccati–Bessel functions having the form of infinite series. Subscript n is a sequential integer index, and primes denote derivatives. The computation of these Mie coefficients is a lengthy process, as the infinite series converge very slowly, particularly for the larger particles. For instance, more than 200 terms of the series are required to obtain adequate convergence for particles of 10 μm radius or greater. Since it is necessary to use a_n and b_n for each value of n, the simple storage of the computed values is a problem in itself for most computational facilities.

The angular dependence of the scattered radiation field is introduced through the relations

$$\pi_n(\cos\Theta) = \frac{1}{\sin\Theta} P_n^1(\cos\Theta)$$

$$\tau_n(\cos\Theta) = \frac{dP_n^1(\cos\Theta)}{d(\cos\Theta)}$$

(7.21)

Once a_n, b_n, π_n, and τ_n are determined for a sufficient number of values of n for convergence of the series, we can immediately determine the amplitude functions for the scattered radiation field. They are given by

$$S_1(\Theta) = \sum_{n=1}^{\infty} \frac{2n+1}{n(n+1)} \left[a_n \pi_n(\cos \Theta) + b_n \tau_n(\cos \Theta) \right]$$

$$S_2(\Theta) = \sum_{n=1}^{\infty} \frac{2n+1}{n(n+1)} \left[b_n \pi_n(\cos \Theta) + a_n \tau_n(\cos \Theta) \right]$$

(7.22)

Finally, the intensity vector of the scattered radiation has the components

$$i_1(\Theta) = | S_1(\Theta) |^2$$

$$i_2(\Theta) = | S_2(\Theta) |^2$$

(7.23)

plus a phase difference $\delta(\Theta)$ between them. For the scattering of unpolarized incident radiation, there is no permanent phase relation between i_1 and i_2, in which case the scattered field is defined completely by the magnitudes of these two components. For the scattering of polarized incident light, however, there is a phase difference between them, which is the same as saying that some elliptic polarization is introduced by the scattering process. Chu and Churchill (1955) have derived expressions for computing $\delta(\Theta)$.

Other quantities of interest given in terms of the above relations are the scattering cross section σ_{sca} given by

$$\sigma_{\text{sca}} = \frac{1}{x^2} \int_0^{\pi} \left[(i_1(\Theta) + i_2(\Theta)) \right] \sin \Theta \, d\Theta$$

(7.24)

and the extinction cross section

$$\sigma_{\text{ext}} = \frac{2}{x^2} \sum_{n=1}^{\infty} (2n+1) \, \text{Re}(a_n + b_n)$$

(7.25)

As mentioned above, the process of computing the various quantities involved in the Mie theory and applying them to scattering in realistic models of planetary atmospheres is relatively demanding in computer time and storage capacity. Fortunately, a number of authors have developed tables of the functions, which can be applied to particular problems. Examples of the tables are those of Lowan (1949), Penndorf (1963), Zelmanovich and Shifrin (1968, 1971), and Deirmendjian (1969). The polarization and intensity fields of radiation scattered by a water droplet have been studied in great detail by Dave (1969).

The above relations apply to the scattering of radiation by a single particle. This, of course, is a long way from the scattering problem in a real atmosphere containing particles of varying sizes, numbers, and physical characteristics, which themselves are far from constant in space and time.

The gap between single-particle scattering theory and an explanation of the intensity and polarization fields in a real atmosphere is covered in three steps, on each of which a large literature has been built up. Unfortunately space limitations here prohibit all but the most cursory description of the steps required. The details are available in standard references, such as van de Hulst (1957), Kerker (1963), Deirmendjian (1969), and others, but the whole subject is still an area of active research.

The first step is the generalization of the theory of scattering by a single particle to the case of a unit volume of a polydisperse (many-sizes) medium. This requires integration of the relations over the size-frequency distribution of the particles, each size being weighted according to its fraction of the total population. In practice, the integration is normally approximated by a summation over a limited number of discrete size intervals. The procedure is relatively straightforward and presents no particular problems.

The second step toward reality is larger and more difficult to negotiate. It consists of the application of primary-scattering theory to a polydisperse medium in which the once-scattered radiation may be subjected to secondary, tertiary, and higher-order scattering processes. This is an extremely difficult problem for the general case, and it has been only in the last decade or so that significant progress has been made on it.

At the present time there are the following five main methods by which the problem of multiple scattering has been attacked. The principal developers of the methods are indicated in parenthesis.

1. *Method of discrete ordinates* (Chandrasekhar and Yamamoto). This method involves the solution of a number of simultaneous linear differential equations. Although it has wide applicability to atmospheric problems, the difficulty in determining the constants of integration has decreased its popularity for actual use.

2. *Doubling method* (van de Hulst and Hansen). This ingenious method achieves the very attractive feature of economy in computer time. Although in its primitive form, the method is applicable only to a homogeneous atmosphere, the nonhomogeneous case can be handled by successive application to any number of homogeneous layers.

3. *Matrix method* (Twomey, Jacobowitz, and Howell). This is similar in many respects to the doubling method. The matrix refers to a discrete set of incident and scattering angles. The restriction to a homogeneous atmosphere may be avoided in the same way as for the doubling method.

4. *Iterative method* (Dave; Herman; Irvine). The iterative method is the most direct solution of the transfer equation, and it is widely used. Its principal disadvantage is the large amount of computer time required, particularly when polarization is taken into account.

5. *Monte Carlo method* (Marchuk and Mihailov; Collins and Wells; Plass and Kattawar). This is a powerful method which has been applied to several problems in radiative transfer. It is unique among the methods listed in that it is applicable to nonplane geometries and to cases of horizontal inhomogeneities. Its principal disadvantages are the large demand on computer facilities and the fact that the results are subject to significant statistical variations which make interpretation difficult.

The third step in applying scattering theory to real atmospheres is the confirmation of theoretical results by measurements of the radiation fields. The most difficult aspect of this problem is in characterizing the atmospheric aerosols which exist at any given time and in making adequate measurements of the radiation fields themselves. Both of these aspects continue to present formidable problems to atmospheric scientists.

A number of authors have computed the polarization characteristics of the light reflected and transmitted by model atmospheres containing various types and amounts of aerosols. Typical results of such computations

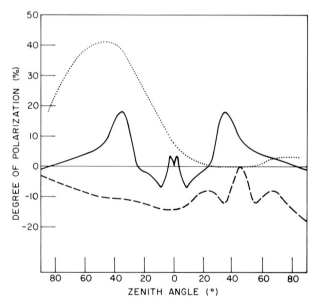

Fig. 7.8 Degree of polarization as a function of zenith angle in the principal plane as computed for sunlight reflected from water droplet clouds (solid and dashed curves) and for sunlight transmitted by a turbid atmosphere (dotted curve). ———: $\lambda = 1.2\ \mu$m, $\tau = 4, \theta_0 = 0°$; (after Hansen, 1971); - - - - - : $\lambda = 3.4\ \mu$m, $\tau = 6, \theta_0 = 45°$ (after Hansen, 1971); ·······: $\lambda = 0.50\ \mu$m, $\tau^R = 0.145$, $\tau^A = 0.23$, $\theta_0 = 49.5°$, $A = 0°$ (after Herman *et al.*, 1971).

are shown in Fig. 7.8. The solid and dashed curves are from the computations of Hansen (1971) for sunlight reflected back to space from the top of water droplet clouds having a size-frequency distribution conforming to the modified γ distribution. The solid curve is for optical thickness $\tau = 4$, $\theta_0 = 0°$, and $\lambda = 1.2$ μm, and the dashed curve is for $\tau = 6$, $\theta_0 = 45°$, and $\lambda = 3.4$ μm. In both cases the polarization is less than 20% everywhere, and for the longer wavelength the sign is consistently negative. For $\lambda = 1.2$ μm, however, the curve shows four neutral points, including the one at $\theta = 0$, in each half of the principal plane. Other computations show a similarly strong dependence on cloud properties and wavelength of the radiation.

Results from the computations of Herman *et al.* (1971) for the case of sunlight of $\lambda = 0.50$ μm transmitted through a turbid atmosphere with both Rayleigh and aerosol components are shown by the dotted curve of Fig. 7.8. A Junge size distribution of slope $\nu^* = -2.5$ was assumed for the aerosol. The general configuration is similar in many respects to that for a pure Rayleigh atmosphere, but the magnitude of the maximum is considerably reduced and the negative branch of the polarization curve is essentially eliminated by aerosol effects. Although the details vary, these results are generally representative of those for other wavelengths and atmospheric optical properties.

7.4 MEASUREMENTS OF SKYLIGHT POLARIZATION

Measurements of the polarization of light in the atmosphere have been pursued for two main purposes. First, the data are useful in assessing the validity of theoretical approximations in the radiative transfer problem, and second, the polarization field has been found to be a sensitive indicator of atmospheric turbidity. Thus measurement programs have been increased during the last decade or so, as a response to the availability of greater numbers of radiative transfer calculations and because of the desire to detect changes of the particulate loading of the atmosphere. As in many fields, however, the measurements of polarization have not kept pace with the needs for accurate data.

The existing observations of skylight polarization, including the dependence of the magnitude of the maximum and the positions of the neutral points on parameters such as atmospheric turbidity and surface reflectance, have recently been summarized by Coulson (1974). Basically, it is observed that the maximum polarization decreases with increasing turbidity of the atmosphere and with increasing surface albedo. The behavior of the neutral points is somewhat erratic, but the Babinet and Brewster points as observed in the blue and ultraviolet regions of the spectrum generally move toward the Sun, and the Arago point toward the anti-Sun, with increasing turbidity. In the red and near-infrared regions, the Babinet and Brewster points tend to disappear entirely, even for relatively low values of atmospheric turbidity.

The general characteristics of the polarization field as observed in a cloudless and moderately hazy atmosphere can be seen from Fig. 7.9. The data were taken on July 25, 1973 at a rural location near Davis, California. Although the horizontal visibility was above 35 km during the period of observations, the haze extended through a deep layer of the atmosphere. Effects of the haze particles were sufficient to decrease the maximum polarization by about 15% at 0.32 μm and 10% at 0.50 μm from its values for a clear atmosphere (Coulson, 1974). Aerosol effects must have dominated the polarization field at 0.70 μm and beyond, as the maximum polarization value progressively decreased with increasing wavelength above about 0.55 μm. This same characteristic of the maximum polarization being highest between 0.50 and 0.60 μm and decreasing at longer and shorter wavelengths was also observed under very clean atmospheric condi-

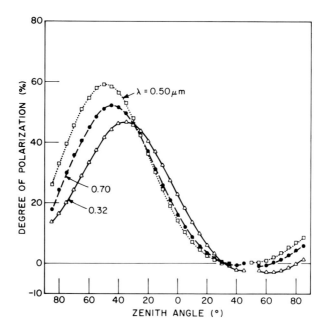

Fig. 7.9 Degree of polarization of radiation as a function of zenith angle in the principal plane as observed in three different wavelengths at Davis, California for moderately hazy atmospheric conditions.

tions at an altitude of 3670 m at the Mauna Loa Observatory in Hawaii (unpublished data).

The distribution of skylight polarization under conditions of moderate to heavy air pollution is shown for a wavelength of $\lambda = 0.32$ μm in Fig. 7.10 for eight different solar elevations (Coulson, 1971). Not only is the maximum polarization decreased by a factor of 2 or more as the pollution density builds up, but also the negative polarization in the vicinity of the Sun is decreased so strongly that the neutral points essentially disappear.

The positions of the neutral points, particularly the Babinet and Arago points, have been studied more extensively than any other feature of the polarization field. The main reasons for this are that they are quite easy to observe visually and they have been found to be sensitive indicators of atmospheric turbidity (Neuberger, 1950). The Brewster point is difficult to observe by eye, so most of the measurements of Brewster point positions have been accumulated since the advent of electronic polarimeters.

Observations of neutral point positions were made by visual means, mostly "white" light, by Brewster (1864), Cornu (1884), Dorno (1919), Neuberger (1950), and others.

The existence of volcanic ash in the atmosphere produces large changes

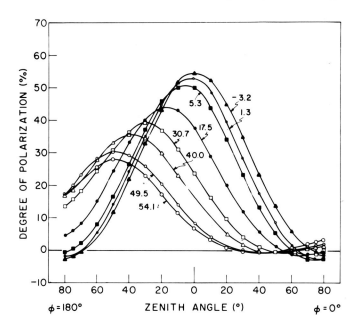

Fig. 7.10 Degree of polarization of radiation as a function of zenith angle in the principal plane as observed at eight different sun elevations under conditions of moderate to heavy air pollution in Los Angeles ($\lambda = 0.32$ μm) (after Coulson, 1971).

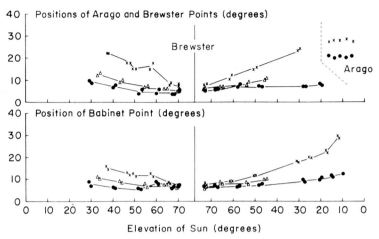

Fig. 7.11 Observed position of the neutral points as a function of Sun elevation for three different wavelengths in relatively clear atmospheric conditions in Los Angeles (○, $\lambda = 0.603 \ \mu$m; △, $\lambda = 0.512 \ \mu$m; ✕, $\lambda = 0.404 \ \mu$m) (after Holzworth and Rao, 1965).

of neutral point behavior, which appears to be unique to volcanic ash. Cornu (1884) observed neutral points to be situated symmetrically on each side of the Sun and anti-Sun after the eruption of Krakatao, and the positions of the Babinet and Arago points were strongly shifted after the eruptions of both Krakatao (1883) and Mt. Katmai (1912). The Babinet point appears to be particularly sensitive to volcanic ash, undergoing at least a $+16°$ shift of position at a Sun elevation of 4°, which changes to a $-2°$ shift when the Sun is 4° below the horizon. The Arago point is less sensitive, but it also shows some anomalous effects produced by volcanic ash.

Series of measurements of neutral point positions at different wavelengths have been made with electronic polarimeters by Sekera *et al.* (1955), Holzworth and Rao (1965), Coulson (1971), and others. Typical locations of the points, as observed at three visible wavelengths by Holzworth and Rao (1965), are plotted as a function of Sun elevation in Fig. 7.11. The measurements were taken under moderate visibilities and clear skies in the Los Angeles area. The shift of neutral point positions toward the Sun (or anti-Sun for the Arago point) with increasing wavelength is clearly seen in the data. This shift is a result of decreasing multiple scattering with increasing wavelength, and corresponds roughly to that predicted by theory.

7.5 INSTRUMENTS FOR MEASURING POLARIZATION

As mentioned above, the type of polarization of most interest in the atmosphere is linear polarization. The degree of elliptic or circular polarization of scattered sunlight is so small as to be at about the limit of detectability, although Rozenberg and his colleagues have studied the field of elliptic polarization on illumination of the atmosphere by an artificial source at night (Rozenberg and Gorchakov, 1967; Rozenberg, 1968).

Polarimeters used for measuring skylight polarization are conveniently divided into visual and electronic types. They will be discussed separately.

7.5.1   Visual Polarimeters

Before about 1930, polarization was measured entirely by visual means, and a number of different visual polarimeters were developed. Glazebrook (1923) lists nearly a dozen different polarimeters or variations thereof, although the majority of them were developed for measuring the optical rotation properties of different chemical solutions.

Visual polarimeters have now become largely obsolete due to the development of electronic type instruments, but they continued to be used for significant observations through the decade of the 1950's (Neuberger, 1950; Pyaskovskaya–Fesenkova, 1958, 1960).

The Cornu-Martens Polarimeter   The most advanced of the visual instruments for skylight measurements was the Martens (1900) version of the Cornu (1890) polarimeter. In this instrument, the incident light passes successively through Wollaston and Nicol prisms before being presented for viewing at a split-field eyepiece. The prisms are rotatable with respect to each other, and as a unit. With the Nicol fixed at 45° to the principal axis of the Wollaston, the unit is rotated until the two halves of the viewed field are of equal intensity. At this point, the plane of polarization of the incident light is inclined 45° from the principal axis of the Wollaston. If the Wollaston is then rotated by 45°, one half transmits the intensity component parallel, and the other half that normal, to the plane of polarization. If the Wollaston is then kept fixed and the Nicol prism is rotated until the two halves of the field again have the same intensity, the degree of polarization P is then given by the relation $P = \cos 2\omega$, where ω is the angle between the principal planes of the two prisms.

The precision obtainable by the Martens polarimeter is somewhat dependent on the magnitude of P, but careful measurements can yield a precision of the order of $\pm 1\%$ if the incident light is sufficiently bright to properly illuminate the two halves of the split field. This aspect is a problem during twilight operations. Other disadvantages are the relatively long time necessary to make a measurement (about 6 min), the impossibility of measurements in narrow spectral regions or at wavelengths outside the visible spectrum, and operator fatigue during long measurement series.

The Savart Polariscope   Visual measurements of the positions of the neutral points are most conveniently made by means of the Savart polariscope. The basis of this device is a Nicol or other analyzing prism in combination with a Savart plate constructed from a quartz crystal. The plate is made by cutting two identical thin sections oriented at an angle of 45° from the optical axis of the crystal, rotating one section 90° with respect to the other, and cementing the two together. Light passing through the resulting plate and then through the analyzer oriented at 45° to each of the axes of the plate forms light and dark fringes, the central fringe being bright or dark if the plane of polarization is parallel or normal, respectively, to the plane of transmission of the plate. In case the light is unpolarized, no fringes appear.

In order to measure the neutral point positions, the device is mounted on a theodolite or other mechanism for measuring angles, and oriented so that the fringes are vertical when the viewed direction is in the plane of the Sun's vertical. The position of a neutral point can then be determined by observing the angle at which the fringes are interrupted, with dark above and bright below, in the field of view of the instrument. This is easily seen if the illumination is sufficiently intense and the degree of polarization increases moderately strongly on each side of the neutral point. Polarization magnitudes of $\pm 1\%$ can easily be detected by the appearance of the fringes.

7.5.2   Electronic Polarimeters

The polarimeters which have been developed since the advent of electronics, photomultiplier tubes, computers, and other components of modern instrumentation systems have been of the electronic type. There are many and powerful reasons for the adoption of the new techniques over the traditional visual methods of polarimetry. Measurements can be extended from the visible into the ultraviolet and infrared spectral regions, and the great sensitivity attainable permits measurements in very narrow regions of the spectrum and small solid angles. Measurements can be made very rapidly and without the necessity of an observer laboriously peering into an instrument and estimating intensities by eye. Data recording and processing may be done automatically by digital techniques, thereby greatly decreasing the cost involved and eliminating the introduction of human errors into the data. The accuracy attainable is much higher in modern techniques than was possible by the older visual methods, and the measurement of low degrees of polarization presents no problem. The great difficulty that Brewster experienced in searching for the neutral point named for him would have been eliminated had he been blessed with the electronic polarimeters available today.

The main disadvantages of electronic type polarimeters are their greater initial cost and their complexity. The greater initial cost, however, is rapidly offset by the saving of manpower in reducing the data obtained if a series of measurements is involved. The greater complexity is not as much of a problem now as it was a decade or two ago because of the advances which have been achieved in electronics, but electronic instruments still require frequent calibration for reliable results.

Probably the first electronic polarimeter was that of Sekera (1935), in which the light transmitted by a rotating Nicol prism was sensed by a photovoltaic cell. The main problems encountered in this first attempt were noise in the electronic system, a long time constant of the recording galvanometer, and the fact that the output of the photocell was sensitive to the orientation of the plane of transmission of the polarizer. As will be seen below, all of these difficulties have been overcome in more modern types of electronic polarimeters.

Before looking at specific examples of electronic polarimeters, a number of which have been developed, it is well to look at the theory on which they are based. As shown by Mueller (1948) and given in detail by Shurcliff (1962), it is possible to represent the action of an optical device, be it retarder, polarizer, or other component, by a 16-element matrix, now called a Mueller matrix. One can set up a train of Mueller matrices, one for each optical component, and use ordinary rules of matrix manipulation to determine the Stokes vector of the transmitted light for a given type of incident light. The Mueller matrix for an ideal linear polarizer is (Shurcliff, 1962)

$$\mathbf{P} = \tfrac{1}{2}\begin{bmatrix} 1 & C_2 & S_2 & 0 \\ C_2 & C_2{}^2 & C_2S_2 & 0 \\ S_2 & C_2S_2 & S_2{}^2 & 0 \\ 0 & 0 & 0 & 0 \end{bmatrix} \tag{7.26}$$

and that for an ideal quarter-wave retardation plate is

$$\mathbf{R} = \begin{bmatrix} 1 & 0 & 0 & 0 \\ 0 & C_2{}^2 & C_2S_2 & -S_2 \\ 0 & C_2S_2 & S_2{}^2 & C_2 \\ 0 & S_2 & -C_2 & 0 \end{bmatrix} \tag{7.27}$$

Here C_2 and S_2 are, respectively, $\cos 2\xi$ and $\sin 2\xi$ for the polarizer and $\cos 2\psi$ and $\sin 2\psi$ for the retardation plate, ξ and ψ being, respectively, the

angles between the reference plane and plane of transmission for the polarizer and between the reference plane and the fast axis of the retardation
plate. By using these matrices as successive operators on the incident light
expressed as the Stokes vector, one can arrive at the Stokes parameters of
the transmitted light. Conversely, one can measure the Stokes parameters
of the transmitted light and determine therefrom the Stokes vector of the
incident light. This latter process is the basis of the electronic polarimeters
described below.

Polarimeters with Rotating Retardation Plate and Fixed Analyzer Two of
the polarimeters which have been developed by Sekera and his collaborators,
(Sekera *et al.*, 1955; Sekera, Rao, and Dibble, 1963) have utilized a rotating
retardation plate in front of a fixed analyzer, of the general type shown
schematically in Fig. 7.12. If light of intensity I and linear polarization P
is incident on such a system, the intensity I_a leaving the analyzer is given
(Sekera *et al.*, 1955) by

$$I_a = I + Q \sec 2\chi [\cos^2(\delta/2) \cos 2(\chi - \psi) + \sin^2(\delta/2)$$
$$\times \cos(4\xi - 2\psi - 2\chi)] + V \sin \delta \sin(\xi - \psi) \quad (7.28)$$

where δ is the retardation of the plate, and the angles between the reference
plane and the fast axis of the retardation plate, plane of transmission of the
analyzer, and the plane of polarization of the incident light are denoted by
ξ, ψ, and χ, respectively. If the retardation plate is rotated around the
optical axis with angular frequency ω, the intensity leaving the analyzer is

$$I_a = I\{1 + P_L[\cos^2(\delta/2) \cos 2(\chi - \psi) + \sin^2(\delta/2)$$
$$\times \cos(4\omega t - 2\psi - 2\chi)] + V \sin \delta \sin(2\omega t - 2\psi)\} \quad (7.29)$$

Here P_L is the degree of linear polarization of the incident light defined by

$$P_L = (Q^2 + U^2)^{1/2}/I \quad (7.30)$$

The signal from an electronic instrument based on this optical system

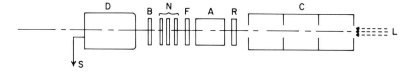

Fig. 7.12 Schematic diagram of the optical system of a polarimeter using a rotating
retardation plate and fixed analyzer. L: incident light; C: collimator tube; R: rotating
retardation plate; A: fixed analyzer; F: color filter; N: neutral density filters; B: Babinet
compensator; D: detector (photomultiplier tube); S: output signal.

is in the form of a composite voltage signal consisting of the following three components:

1. A steady component v_0 given by

$$v_0 = KI[1 + P_L \cos^2(\delta/2) \cos 2(\chi - \psi)] \qquad (7.31)$$

where K is the calibration factor of the instrument. It is assumed that K takes account of any neutral density filters or voltage control circuits for controlling instrument sensitivity.

2. A second harmonic v_2 expressed as

$$v_2 = KIV \sin \delta \sin(2\omega t - 2\psi) \qquad (7.32)$$

The amplitude of this harmonic is used to determine V, and thence the ellipticity β from the relation $\sin 2\beta = V/I$. It is assumed that I is determined from the other two components of the signal. The phase of the second harmonic can be used to determine ψ if it is not known otherwise.

3. A fourth harmonic v_4 given by the relation

$$v_4 = KIP_L \sin^2(\delta/2) \cos(4\omega t - 2\psi - 2\chi) \qquad (7.33)$$

The amplitude of v_4 can be combined with v_0 to determine both I and P_L, and the phase of v_4 gives the angle χ of the plane of polarization.

This completes the analysis of the incident light, the Stokes vector of which is $\{I, Q, U, V\}$. I is found from v_0 and the amplitude of v_4. Q and U are determined from a combination of Eq. (7.30) and the relation $U = Q \tan 2\chi$, where P_L is found from v_0 and the amplitude of v_4 and χ is found from the phase of v_4. Finally V is determined from the amplitude of v_2.

The polarimeter of Sekera *et al.* (1955) utilized a rotating quarter-wave plate in combination with a fixed Glan–Thompson analyzer. It was a dual-channel instrument based on the use of photomultiplier tubes. Each of the two channels was fitted with two spectral filters of the absorption type, the wavelengths of peak transmission for the four filters being 0.365, 0.460, 0.515, and 0.625 μm. One quarter-wave plate was used in each channel, the effective wavelength of each being midway between the peak transmissions of the two filters of the channel. The dynamic range necessary to cover the large changes of intensity of skylight was attained by varying the voltage applied to the photomultiplier in such fashion as to keep the anode current at a low and constant value. The impressed voltage was then calibrated in terms of intensity, and its value was recorded during the measurement operation. By this means the current could be kept low enough to minimize fatigue effects in the photomultiplier. The amplitudes of the second and fourth harmonics of the signal were recorded separately by means of tuned amplifiers, and the phase of the fourth harmonic was sensed by a phase detector and recorded.

Although this first instrument was used for many measurements of skylight, and another slightly improved three-channel version of it was later developed, there were several difficulties with it. The theoretical possibility of separating the various types of signals for recording was never satisfactorily accomplished, and the time constant of the voltage control system was too long to faithfully reproduce rapid intensity changes. An even greater problem was apparent unequal transmission along the fast and slow axes of the retardation plates. This resulted in the amplitude and phase of the second harmonic being dependent, not only on the ellipticity of the light, but also on the intensity, degree of linear polarization, and orientation of the plane of polarization as well. The problem defied all efforts at solution and essentially eliminated the possibility of reliable measurements of the always small values of elliptic polarization in the atmosphere.

A second polarimeter of Sekera and collaborators (Sekera *et al.*, 1963) was based also on the system of rotating retardation plate with fixed analyzer, but the improvements over the previous instrument were substantial. A large dynamic range was attained by use of an electronically positioned, neutral density wedge for controlling the intensity of light to the photomultiplier tube, the position of the wedge being calibrated and recorded as the intensity observable. The decision to neglect the small amount of elliptic polarization in the skylight permitted the use of half-wave plates instead of the previously used quarter-wave plates, thereby maximizing sensitivity to the plane-polarized component. Four matched pairs of interference filters and half-wave plates centered at wavelengths 0.3325, 0.410, 0.510, and 0.610 μm were compatible with the response characteristics (between S-11 and S-13) of the antimony–cesium photocathode of the photomultiplier tube. Other significant differences in this from the previous instrument are the use of one instead of two channels, the provision of battery operation and signal telemetry for airborne application, and the use of a Glan prism instead of a Glan–Thompson prism to obtain maximum sensitivity in the ultraviolet. The instrument was sent aloft by balloon on two different occasions in 1964 and 1965 to measure the polarization of light directed upward at high altitudes, but unfortunately only partial success was achieved. Later flights with a different instrument, described below, yielded much better results.

One serious difficulty with both of the above polarimeters can be seen from Eq. (7.29). The degree of linear polarization is a factor in the second and third terms, both of which involve the variable and generally unknown angle $\chi - \psi$ between the plane of polarization of the incident light and plane of transmission of the analyzer. For perfectly monochromatic light and a half-wave plate, the factor $\cos^2(\delta/2)$ would vanish, in which case P_L could be determined only from the amplitude of the fourth harmonic. Unfortunately, the finite bandwidth of an optical filter eliminates this possibility. The only other way to make the second term vanish is to set $|\chi - \psi| = 45°$. In both of the polarimeters described above the analyzer has been fixed at $\psi = 45°$, which means that χ must be either $0°$ or $90°$. In the (clear) atmosphere, $\chi = 0°$ or $90°$ only in the vertical plane through the azimuth of the Sun. Thus measurements with the above polarimeters have been generally confined to this single plane. The inability to interpret instrument signals at other azimuths was largely responsible for the abandonment of this system for balloon-borne measurements.

Polarimeters with Rotating Analyzer The simplest type of polarimeter is based on an optical system similar to that shown in Fig. 7.12, except that the retardation plate is omitted and the analyzer is rotated around the optical axis. For this case, the intensity $I(\psi)$ falling on the detector at angle ψ of the plane of transmission of the analyzer is, aside from instrumental constants, given by

$$I(\psi) = I_e \cos^2 \psi + I_r \sin^2 \psi + \tfrac{1}{2} U \sin 2\psi \qquad (7.34)$$

Thus we have three unknowns (I_e, I_r, U), which can theoretically be determined by three measurements of I at properly chosen values of ψ, from which the total intensity of I, degree of plane polarization P, and orientation of the plane of polarization χ, can be determined from their definitions given in Eqs. (7.1)–(7.3).

It is well known, however, that the accuracy of the results can be increased by increasing the number of independent measurements. In fact, if n is the number of independent measurements, all of comparable accuracy, then the standard deviation of the various parameters is proportional to $(n)^{-1/2}$. In order to take advantage of this characteristic, Coulson (1969b) increased the number of measurements made in one rotation of the analyzer to 12, which were spaced at intervals of ψ of $30°$. Then the 12 values of intensity $I(\psi_i)$ producing the corresponding signals v_i, $i = 1, 12$ were resolved into Fourier components given by the series

$$I(\psi) = p_0 + p_1 \cos \psi + p_2 \cos 2\psi + \cdots + q_1 \sin \psi + q_2 \sin 2\psi + \cdots$$

$$(7.35)$$

where the mean is

$$p_0 = \frac{1}{12} \sum_{i=1}^{12} v_i \qquad (7.36)$$

and the coefficients for the kth harmonic $(k \leq 5)$ are

$$p_k = \frac{1}{6} \sum_{i=1}^{12} v_i \cos(ik\alpha)$$

$$q_k = \frac{1}{6} \sum_{i=1}^{12} v_i \sin(ik\alpha)$$

(7.37)

The Stokes parameters are given in terms of these coefficients by:

$$I = 2p_0 \qquad Q = 2p_2 \qquad U = 2q_2 \tag{7.38}$$

from whence the degree and plane of polarization are

$$P = (p_2{}^2 + q_2{}^2)^{1/2}/p_0$$

$$\chi = \tfrac{1}{2} \tan^{-1}(q_2/p_2)$$

(7.39)

The physical reason that all of the polarization information is contained in the second harmonic is that for given conditions the same amount of light is transmitted by the prism oriented at $\psi + \pi$ as at orientation ψ.

The simple rotating analyzer system has been used by many workers, including Chen *et al.* (1967), Bullrich (1964), Fernald *et al.* (1969), Weinberg (1964), Coulson *et al.* (1965), Coulson and Walraven (1972), and others. Although the optical systems of the various devices vary in details, the configuration is generally similar to that shown in Fig. 7.12, but with the retardation plate, and in some cases the Babinet compensator, omitted.

In astronomical observations the telescope takes the place of the collimator tube of Fig. 7.12. As Hiltner (1962) points out, however, extreme care must be taken in astronomical measurements to assure that polarization effects introduced by the telescope are properly accounted for. Because of this problem, Hall and Mikesell (1950) inserted a depolarizer and calibrator in sequence in front of the rotating analyzer to detect spurious polarization signals. They measure the polarization of the incoming light by matching its signal with the signal introduced by the calibrator set to give a known degree of polarization. Because of the problem of spurious polarization effects in the system itself, experimenters working in the atmospheric radiation regime have mainly avoided entrance optics, although some (e.g., Bullrich, 1964) use a simple lens at the entrance aperture of the instrument to increase its light gathering capacity.

The simplest and most inexpensive type of analyzer is the plastic sheet polarizer usually designated as polaroid. It has been used, apparently quite satisfactorily, in several instruments (e.g., Weinberg, 1964; Fernald *et al.*, Bullrich, 1964). However, because of the low polarizing efficiency of this material at short wavelengths and its tendency to become distortetd, the use of a Glan–Thompson prism, Glan prism, or other type of birefringent crystal has normally been preferred. The depolarizer, shown in Fig. 7.12 in front of the photomultiplier tube, is necessary for side-window tubes, as the response of such tubes is usually a function of the orientation of the plane of polarization of the light falling on them. Most modern end-window photomultipliers are not so subject to the difficulty but even the end-window tubes vary in this respect.

Various methods have been used for recording and processing of the output signal of electronic polarimeters. In most instances analog techniques have been employed, the analog signal being either recorded directly on a strip chart or analog tape recorder. The disadvantage of this approach is that further processing, which may be laborious and time consuming, is required to yield the physical quantities of interest. A more direct approach is to transfer the signal to digital form for recording on either punched paper or digital magnetic tape. This facilitates eventual processing by electronic computer techniques and greatly speeds up the process of data reduction.

A completely digitalized system has been developed by Coulson and Walraven (1972) for routine measurements of skylight polarization. Photon counting techniques are used in this system to gain the double advantage of high precision in the measurements and a completely digitalized system which can be operated entirely by computer. The mini-computer, which is an integral part of the system, not only performs all of the signal processing and computations necessary to yield a real-time print-out of the intensity, degree of polarization, and orientation of the plane of polarization, but it also controls the operation of the instrument itself.

A photograph of this instrument, together with the mini-computer by which it is operated, is shown in Fig. 7.13. Each of the two channels is fitted with four interference filters, thereby providing the possibility of making measurements in any of eight different spectral ranges. The filters in use at present are of the interference type with bandwidths of about 0.01 μm and are centered at 0.32, 0.365, 0.40, 0.50, 0.60, 0.70, 0.80, and 0.90 μm. The goniometer-type mount is controlled by digitally actuated stepping motors, by means of which any desired scan pattern may be achieved by appropriate computer programs.

The extremely rapid advances presently being made in high-speed electronics, in combination with the economy and convenience to be gained by direct signal processing and the basic simplicity of the rotating polarizer

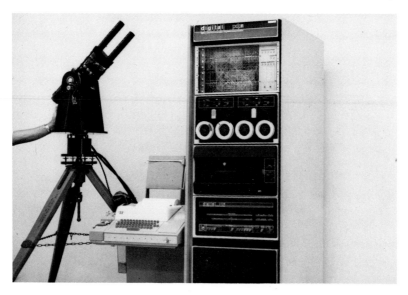

Fig. 7.13 Photograph of the dual-channel electronic polarimeter developed by Coulson and Walraven (1972) for measuring the polarization of skylight. The mini-computer is used for controlling the operation of the instrument and for data processing and printout of the results on a real-time basis.

system, would seem to indicate the direction that polarimeter development is likely to take in the future.

REFERENCES

Ambartsumian, V. A. (1943). On the problem of the diffuse reflection of light by a turbid medium. *Dokl. Akad. Nauk USSR* **38,** 229–257.

Ambartsumian, V. A. (1944). Diffusion of light through a scattering medium of large optical thickness. *Dokl. Akad. Nauk USSR* **43,** No. 3.

Born, M., and Wolf, E. (1959). "Principles of Optics." Pergamon, Oxford.

Brewster, D. (1864). Observations of the polarisation of the atmosphere, made at St. Andrews in 1841, 1842, 1843, 1844, and 1845. *Trans. Roy. Soc. Edin.* **23,** 211–240.

Bullrich, K. (1964). Scattered radiation in the atmosphere and the natural aerosol. *Advan. Geophys.* **10,** 99–260.

Bullrich, K., Eiden, R., and Nowak, W. (1966). Sky radiation, polarization, and twilight radiation in Greenland. *Pure Appl. Geophys.* **66,** 220–242.

Chandrasekhar, S. (1950). "Radiative Transfer." Oxford Univ. Press (Clarendon), London and New York.

Chandrasekhar, S., and Elbert, D. D. (1951). Polarization of the sunlit sky. *Nature* (*London*) **167**, 51–54.

Chen, H. S., Rao, C. R. N., Sekera, Z. (1967). Investigations of the polarization of light reflected by natural surfaces. Sci. Rept. No. 2, Contr. AF 19 (628)-3850, University of California, Los Angeles, California.

Chandrasekhar, S., and Elbert, D. D. (1954). Illumination and polarization of the sunlit sky on Rayleigh scattering. *Trans. Amer. Phil. Soc.* **44**, 643–728.

Chen, H. S., and Rao, C. R. N. (1968). Polarization of light reflected by some natural surfaces. *Brit. J. Appl. Phys.* **1**, (2), 1191–1200.

Cornu, A. (1884). Observations relatives à la couronne visible actuellement autour due Soleil. *Comp. Rend. Acad. Sci., Paris* **99**, 488–493.

Chu, C. M., and Churchill, S. W. (1955). Representation of the angular distribution of radiation scattered by a spherical particle. *J. Opt. Soc. Amer.* **45**, 958–962.

Cornu, A. (1890). Sur l'application du photopolarimetre a la meteorologie. C.R. Ass. Franc. Av. Sci., Session a Limoges, pp. 267–270.

Coulson, K. L. (1952). Polarization of light in the Sun's vertical. Sci. Rept. No. 4, Contr. AF19(122)-239, Univ. Calif. Los Angeles, California.

Coulson, K. L. (1959). Characteristics of solar radiation emerging from the top of a Rayleigh atmosphere. *Planet. Space Sci.* **1**, 265–276.

Coulson, K. L. (1968). Effect of surface reflection on the angular and spectral distribution of skylight. *J. Atmos. Sci.* **25**, 759–770.

Coulson, K. L. (1969a). Polarimetry of Mars. *Appl. Optics* **8**, 1287–1294.

Coulson, K. L. (1969b). Measurements of ultraviolet radiation in a polluted atmosphere. Sci. Rep. No. 1, Grant No. 5 R01 AP 00742-02, Univ. Calif., Davis, California.

Coulson, K. L. (1971). On the solar radiation field in a polluted atmosphere. *J. Quant. Spectrosc. Radiat. Transfer* **11**, 739–755.

Coulson, K. L. (1974). The polarization of light in the environment. *In* "Planets, Stars, and Nebulae Studied with Photopolarimetry" (T. Gehrels, ed.). Univ. of Arizona Press. Tucson, Arizona.

Coulson, K. L., and Walraven, R. L. (1972). A high-precision computer–controlled dual–channel polarizing radiometer. Preprints, Conf. on Atmos. Rad., Fort Collins, Colo., pp. 161–162.

Coulson, K. L., Dave, J. V., and Sekera, Z. (1960). "Tables Related to Radiation Emerging from the Top of a Planetary Atmosphere with Rayleigh Scattering." Univ. of California Press, Berkeley.

Coulson, K. L., Bouricius, G. M., and Gray, E. L. (1965). Optical reflection properties of natural surfaces. *J. Geophys. Res.* **70**, 4601–4611.

Dave, J. V. (1964). Importance of higher order scattering in a molecular atmosphere. *J. Opt. Soc. Amer.* **54**, 307–315.

Dave, J. V. (1969). Scattering of visible light by large water spheres. *Appl. Opt.* **8**, 155–164.

Dave, J. V., and Furukawa, P. M. (1966). Intensity and polarization of the radiation emerging from an optically thick Rayleigh atmosphere. *J. Opt. Soc. Amer.* **56**, 394–400.

deBary, E. (1964). Influence of multiple scattering on the intensity and polarization of diffuse sky radiation. *Appl. Opt.* **3**, 1293–1303.

deBary, E., and Bullrich, K. (1964). Effects of higher order scattering in a molecular atmosphere. *J. Opt. Soc. Amer.* **54**, 1413–1416.

Deirmendjian, D. (1969). "Electromagnetic Scattering on Spherical Polydispersions." Elsevier, Amsterdam.

Deirmendjian, D., and Sekera, Z. (1954). Global radiation resulting from multiple scattering in a Rayleigh atmosphere. *Tellus* **6**, 382–398.

Dorno, C. (1919). Himmelshelligkeit, himmelspolarisation, und sonnenintensitat in Davos (1911 bis 1918). *Meterol. Z.* **36**, 109–124, 181–192.

Fernald, F. G., Herman, B. M., and Curran, R. J. (1969). Some polarization measurements of the reflected sunlight from desert terrain near Tucson, Arizona. *J. Appl. Meteorol.* **8**, 604–609.

Fraser, R. S. (1964). Theoretical investigation: the scattering of light by a planetary atmosphere. Final Rep., Contr. NAS 5-3891, TRW Space Tech. Lab., Redondo Beach, California.

Gehrels, T. (1962). Wavelength dependence of the polarization of the sunlit sky. *J. Opt. Soc. Amer.* **52**, 1164–1173.

Gehrels, T., and Teska, T. M. (1963). The wavelength dependence of polarization. *Appl. Opt.* **2**, 67–77.

Glazebrook, R. (1923). "Dictionary of Applied Physics," Vol. 4. Macmillan, London.

Hall, J. S., and Mikesell, A. H. (1950). Polarization of light. U.S. Naval Observatory Pub. **17**, (1), 1–62.

Hansen, J. E. (1971). Multiple scattering of polarized light in planetary atmospheres, Part II. Sunlight reflected by terrestrial water clouds. *J. Atmos. Sci.* **28**, 1400–1426.

Herman, B. M., Browning, S. R., and Curran, R. J. (1971). The effect of atmospheric aerosols on scattered sunlight. *J. Atmos. Sci.* **28**, 419–428.

Hiltner, W. A. (1962). "Astronomical Techniques." Univ. of Chicago Press. Chicago, Illinois.

Holzworth, G. C., and Rao, C. R. N. (1965). Investigations of the polarization of the sunlit sky. Sci. Rep. No. 1, Contr. AF 19 (628)-3850, Univ. of Calif., Los Angeles, California.

Jensen, C. (1942). Die polarisation des himmelslichtes. *Meteorol. Z.* **43**, 132–140.

Kerker, M. (ed.) (1963). "Electromagnetic Scattering." Pergamon, Oxford.

King, L. V. (1913). On the scattering and absorption of light in gaseous media, with application to the intensity of sky radiation. *Phil. Trans. Roy. Soc. London Ser. A* **212**, 375–433.

Lowan, A. N. (1949). Tables of scattering functions for spherical particles. *Appl. Math. Ser.* 4, Nat. Bur. of Stds., Washington, D.C.

Martens, F. F. (1900). Uber ein neues polarisationsphotometer fur weisses licht. *Phys. Z.* **1**, 299–303.

Mueller, H. (1948). The foundations of optics. *J. Opt. Soc. Amer.* **38**, 661.

Neuberger, H. (1950). Arago's neutral point: A neglected tool in meteorological research. *Bull. Amer. Meteorol. Soc.* **31**, 119–125.

Penndorf, R. (1963). Research on aerosol scattering in the infrared. Final Rep., AFCRL–63-688, Bedford, Mass.

Pyaskovskaya-Fesenkova, E. V. (1958). On scattering and polarization of light in desert conditions. *Dokl. Akad. Nauk USSR*, **123**(6), 1006–1009.

Pyaskovskaya–Fesenkova, E. V. (1960). Some data on the polarization of atmospheric light. *Dokl. Akad. Nauk USSR* **131**(2), 97–299.

Rozenberg, G. V. (1968). Optical investigations of atmospheric aerosols. *Sov. Phys. Usp.* **11**, 353–380.

Rozenberg, G. V., and Gorchakov, G. I. (1967). The degree of ellipticity of the polarization of light scattered by atmospheric air as a tool in the investigation of aerosol microstructures. *Izv. Atmos. Ocean Phys.* **3**, 400–407.

Sekera, Z. (1935). Lichtelektrische registrierung der himmelspolarisation. *Gerlands Beitr. Geophys.* **44**, 157–175.

Sekera, Z.(1950). Polarization of skylight. *In* "Compendium of Meteorology," pp. 79–90. Amer. Meteor. Soc., Boston, Massachusetts.

Sekera, Z. (1956). Recent developments in the study of the polarization of skylight. *Advan. Geophys.* **3**, 43–104.

Sekera, Z. (1957). Polarization of skylight. "Encyclopedia of Physics," Vol. 48, pp. 288–328. Springer Publ., New York.

Sekera, Z., Coulson, K. L., Deirmendjian, D., Fraser, R. S., and Seaman, C. (1955). Investigation of polarization of skylight. Final Rep., Contr. AF 19(122)-239, Univ. of Calif., Los Angeles, California.

Sekera, Z., Rao, C. R. N., and Dibble, D. (1963). Photoelectric skylight polarimeter. *Rev. Sci. Instrum.* **34**, 764–768.

Shurcliff, W. A. (1962). "Polarized Light." Harvard Univ. Press. Cambridge, Massachusetts.

Sobolev, V. V. (1949). On the diffuse reflection and transmission of light by a plane layer of a turbid medium. *Dokl. Akad. Nauk USSR* **69,** 353.

Sobolev, V. V. (1950). "Radiative Transfer in Stellar and Planetary Atmospheres." Gidrometeoizdat, Leningrad, U.S.S.R.

Soret, J. L. (1888). Sur la polarisation atmospherique. *Arch. Sci. Phys. Natur.* **20,** 429–471.

Stone, J. M. (1963). "Radiation and Optics." McGraw–Hill, New York.

Strutt, J. W. (1871). On the light from the sky, its polarisation and colour. *Phil. Mag.* **41,** 107–120, 274–279.

van de Hulst, H. C. (1957). "Light Scattering by Small Particles." Wiley, New York.

Weinberg, J. (1964). The zodiacal light at 5300 A. *Ann. Astrophys.* **27,** 718–738.

Wood, R. W. (1934). "Physical Optics," 3rd ed. Macmillan. New York.

Zelmanovich, I. L., and Shifrin, K. S. (1968). "Tables of Light Scattering," Vol. 3. Gidrometeoizdat, Leningrad.

Zelmanovich, I. L., and Shifrin, K. S. (1971). "Tables of Light Scattering," Vol. 4. Gidrometeoizdat, Leningrad.

CHAPTER EIGHT

Duration of Sunshine

8.1 USEFULNESS OF MEASUREMENTS OF SUNSHINE DURATION

The duration of sunshine, defined as the amount of time the disk of the Sun is not obscured by clouds, is one of the oldest types of radiation measurement, and a number of different sunshine recorders have been devised over the last 140 years. Data on sunshine duration are valuable for two main purposes. First, sunshine duration, or percent of possible sunshine, is one of the primary parameters in characterizing the climate of a given location, the larger the number of hours of bright sunshine the more desirable the climate for most people. The second use for measurements of sunshine duration is in deducing the total flux of solar radiation on a horizontal surface at locations for which no pyranometric measurements are available. Because of the simplicity, convenience, and relatively low cost of sunshine recorders, a much larger number of stations have been established for measuring sunshine duration than for measuring the flux of global radiation; and a number of studies have shown a high correlation between global radiation and hours of sunshine, particularly when the measurements are integrated over periods of several days or a month. For instance, Bennett (1967) supplemented the global radiation measurements from 57 stations in the United States and 27 stations in Canada with radiation values, estimated on the basis of sunshine duration, at 113 United States stations and 54 Canadian stations, thereby increasing the effective number of stations from 84 to 251. The details of the method and the results obtained in these types of estimations are discussed in Section 8.5.

    In order to indicate how sunshine may be important in various contexts, we quote some remarks attributed to the President of the Optical Society of London following a presentation on solar radiation measurements by R. S. Whipple, 11 March 1915 (*Transactions of the Optical Society,* **xv,** 169).

> But he [the President] could never see a sunshine recorder without thinking it had a romance of its own, because in 1789 the French Revolution had broken out, and there was a passage in history which spoke of the influence of the Sun's rays there. He did not know whether Mr. Whipple was acquainted with the episode, but it might be interesting to recall it. They had in those days little scientific toys, and one of them was a brass cannon fired by the Sun as it passed its meridian. One Sunday—he thought it was July 12, 1798—there was a very excited crowd in the Palais Royale, Paris. News had come through that the King had refused to sanction the requisitions of the National Convention, then sitting at Versailles. And while this crowd of malcontents were considering what they should do under the circumstances, the Sun passed its meridian and the cannon went off. A French oculist of the day, who was a most disreputable member of his (the President's) profession, named Jean Paul Marat—(laughter) —was editor of a paper called "Le Voix du Peuple," and in that paper it was recorded what a tremendous sensation was caused in that assembly by the solemn roar of that little gun, how Camille Desmoulins sprang on a table, seized the leaves from the boughs of a tree, and placed them in his hat as a kind of cockade, how all the crowd did likewise, and how, two days afterwards, the Bastille fell. Well, it seemed to him that that little gun, fired by the Sun, was an instance of solar radiation having a marked influence on human progress.

    The quantity measured by a sunshine recorder is the amount of time, usually expressed to the nearest 0.1 hour, in which the direct solar radiation is of sufficient intensity to activate the recorder. Roughly, this corresponds to the amount of time in which the Sun casts a visible shadow. Since the various recorders have different threshold values for activation, however, they do not all give the same results. Furthermore, none, with the possible exception of the Foster sunshine switch, is sensitive enough to respond to the low intensities which exist during the first few minutes just after sunrise or before sunset. Corrections are normally applied for these periods. A rough approximation of the amount of energy in the direct beam necessary to activate most sunshine recorders is 0.12 cal cm^{-2} min^{-1}, although variations by as much as a factor of 2 from that value have been observed for different types of instruments. In an effort to obtain comparable results from the different instruments, the World Meteorological Organization has recommended that all data be reduced to that which would be observed with a designated standard instrument (Campbell–Stokes sunshine recorder) operating under the same conditions. This recommendation is discussed in more detail in the section on the Campbell–Stokes recorder.

Four main types of sunshine recorders have been developed, none of which is completely satisfactory in all respects. Two types, the Campbell–Stokes and Maring–Marvin, sunshine recorders, use the heat of direct solar radiation to activate the instrument. The third type, typified by the Jordan, Pers, and McLeod models, uses the photochemical effect of sunshine on sensitized paper as the observable, while the photoelectric effect resulting from radiation incident on photosensitive cells is the principle used in the Foster sunshine switch and in recorders based on silicon solar cells. These instruments are discussed individually in the following sections.

8.2 EARLY DEVELOPMENT OF SUNSHINE RECORDERS

8.2.1 Jordan Sunshine Recorders

The first two automatic sunshine recorders were built in the 1838–1840 period by T.B. Jordan, mathematician and instrument maker connected with the Cornwall Polytechnic Society, Falmouth, England (Maring, 1898). The sensitivity of silver chloride to light had just been discovered at that time, so Jordan used a paper covered with silver chloride as a recording medium. In the first instrument, the sensitized paper was mounted on a clock-driven drum which was placed behind the top of the mercury column of a standard barometer. The height of the mercury column determined how much of the sensitized paper was exposed to sunlight, and the intensity of the discoloration was a measure of the intensity of the sunlight. Thus the device combined one of the first automatically recording barometers with the first automatic sunshine recorder, and provided data for a study of the relationship of sunshine duration to barometric pressure.

The second automatic sunshine recorder of T. B. Jordan (about 1840) was also based on sensitized paper as the recording medium, but the design was much simplified from the previous type. It was basically two light metal cylinders, one inside the other, with a sensitized paper covering the outside surface of the inside cylinder. The outside cylinder was made to rotate around the stationary inside cylinder at the rate of one revolution every 24 hr, the rotation being produced by use of a clock driven mechanism. The metal of the outside cylinder was pierced by a small hole, through which sunlight entered and exposed the sensitized paper, and the position and rate of rotation of the cylinder was adjusted so that the hole was always subject to direct illumination by the Sun. Movement of the hole with respect to the paper produced a line of discoloration on the paper, the intensity of the discoloration being a function of the intensity of the sunlight. By mounting the outside cylinder on a helical axis, the hole traced a spiral path over the paper so that successive daily sunshine records could be obtained on the same sheet of paper.

The original model had the disadvantage that diffuse as well as direct radiation would produce a trace on the paper, thereby making it necessary to judge the times of direct sunshine by the intensity of the trace. A later and simpler model of a photographic sunshine recorder, developed in 1885 by J. B. Jordan, son of the original inventor, is shown in Fig. 8.1. It required no moving parts, and consisted of a single cylinder in the sides of which were two small holes located 90° apart. The Sun shining through one of the holes in the forenoon and the other in the afternoon produced two traces on the photographic paper which lined the inside of the cylinder. The overlap of the traces near noon was reduced in a later model, by a small rooflike structure which was attached to the top of the cylinder. A later model of the Jordan instrument is somewhat different from the original one in details, but the principle of operation is similar. In this device, the cylinder was mounted as separate half-cylinders, each of which was lined by a sheet of photosensitive paper. The round holes were replaced

Fig. 8.1 Original sunshine recorder of J. B. Jordan (1885).

by small slits in the flat sides which were used to enclose the half-cylinders (Curtis, 1898). The main disadvantage of this model is that the record is on two pieces of paper instead of one as in the original version.

The main advantages of the Jordan sunshine recorder were its simplicity and the fact that a primary permanent record was obtained. However, daily adjustment was required to prevent an overlap of the traces, and special handling of the photographic paper presented operational problems. In addition, the records were subject to more uncertainty than those of the Campbell–Stokes instrument, mainly because of inconsistency in the sensitivity of the photographic paper.

8.2.2 Marvin Sunshine Recorder

C. F. Marvin of the U.S. Weather Bureau developed, also in the 1880's, a sunshine recorder similar in many ways to that of J. B. Jordan (Middleton, 1969). It also consisted of a single, closed metal cylinder, mounted with its longitudinal axis parallel to the axis of the Earth. The interior of the cylinder was fitted with two curved pieces of metal, each of which held a sheet of sensitized paper. The main improvement over Jordan's design was the inclusion of two movable slides, one on each side of the cylinder, through which the entrance holes were pierced. The slides were moved slightly along a graduated scale once each day, thereby allowing complete records for successive days (throughout a period of a month) to be impressed on a single pair of sensitized sheets without overlap of the traces.

8.2.3 Pers Sunshine Recorder

An optical system, consisting of a hemispherical mirror and a focusing lens, cast an image of the Sun on a sheet of photographic paper in this instrument. When properly mounted, with the optical axis of the instrument parallel to the axis of the Earth, the apparent movement of the Sun caused a movement of the Sun's image on the paper, thereby producing a continuous trace during periods of direct sunshine. No adjustments were required once the instrument was mounted, but the photographic paper had to be changed daily to prevent the overlap of successive traces. As with photographic instruments in general, inconsistency of paper sensitivity caused considerable errors in interpretation of the records. The Pers sunshine recorder was never widely accepted.

8.2.4 McLeod Sunshine Recorder

A later development using sensitized paper was McLeod's sunshine recorder, which dates from about 1880. In this recorder a glass sphere was silvered inside and mounted in front of the lens of a camera. The optical axis of the camera was fixed parallel to the rotational axis of the Earth. Light from the Sun was reflected from the sphere and entered the camera, forming an image on the photographic paper. With movement of the Sun the image moved in an arc of a circle on the paper. Interruption of the continuous line occurred when the Sun was obscured by clouds, and even a 1 min obscuration was visible as a weakening of the intensity of the line on the paper.

8.2.5   Dines Sunshine Recorder

The sunshine recorder developed by W. H. Dines (1900) operated on somewhat the same principle as that of the Maring–Marvin sunshine recorder, discussed in Section 8.3.2. In the Dines instrument a blackened thermometer bulb was filled with ether, and a thread of mercury was inserted in the small bore of the thermometer tube. The thermometer was mounted on pivots so as to be approximately balanced in a horizontal position. When subjected to direct sunlight the ether heated up and expanded, thereby moving the thread of mercury. Movement of the mercury caused the thermometer to become unbalanced and to tip over sufficiently for the mercury to complete an electrical circuit, which, in turn, actuated a recorder pen on a clock-driven drum.

As with the Maring–Marvin instrument, the Dines sunshine recorder was subject to effects of ambient temperature fluctuations and to inability to distinguish between direct and diffuse radiation. It was basically a fragile instrument, and a very critical adjustment of the pivot point was required. It was perhaps these shortcomings which prevented its general acceptance.

8.2.6   Pole-Star Recorder

The same principle employed for sunshine recorders was used for the design of a pole-star recorder by E. C. Pickering in 1885 (Covert, 1925). In this instrument, light from the star Polaris was recorded at night on a photographic film which was mounted on a rotating disk. Interruption of the trace on the film indicated an obscuration of the star by clouds, and it was claimed that a rough idea of the thickness of the cloud could be obtained from the density of the trace. The instrument was essentially a small telescope employing a 75-in. focal-length lens mounted in a telescope tube 4 in. in diameter and 87.5 in. long. By an ingenious device, a time scale was exposed alongside the star trail on the film. A clock operated a shutter to prevent exposure of the film during sunlit periods. Although this appears to be the only instrument which has been developed for the measurement of nighttime cloudiness, it was never widely accepted. According to Covert (1925), probably only four were in existence in 1925, one being at Blue Hill, one at the Greenwich Observatory, one at the University of Chicago, and possibly one in Germany.

8.2.7 Campbell Sunshine Recorder

The original "burning" recorder, which was invented by the Scotsman J. F. Campbell in 1853, used a glass ball filled with water as a spherical lens for concentrating the Sun's rays on the inside surface of a white stone bowl of about 4 in. inside diameter (Glazebrook, 1923). The inside of the bowl was engraved with hour lines and painted with an oil paint or varnish. The heat of the concentrated solar rays melted the paint during periods of direct sunshine, but not when the solar disk was covered by clouds. The periods of direct sunshine were read and tabulated daily. Campbell suggested the use of wood for the bowl, and, at a later time, the concentrating lens was mounted in a shallow mahogany bowl of radius such that the surface of the wood always coincided with the focal point of the lens. As the instrument was exposed with the same wooden bowl day after day, a series of grooves was burned into the interior surfaces of the bowl. The bowls were replaced at 6-month intervals, the normal period of use extending from one solstice to the next. The water-filled bulb was replaced by a solid glass sphere in 1857 (Curtis, 1898), but the wooden bowls continued in use for several years (Maring, 1898). A photograph of two of these curious and interesting old records is shown in Fig. 8.2; the original bowls are on display at the Science Museum, London.

Fig. 8.2 Two of the mahogany bowls used in the original burning glass sunshine recorder of Campbell (copyright of the Science Museum, London; reproduced by permission).

8.3 MODERN SUNSHINE RECORDERS

8.3.1 Campbell–Stokes Sunshine Recorder

A significant modification of the Campbell sunshine recorder by Sir G. G. Stokes in 1879 resulted in the Campbell–Stokes sunshine recorder assuming essentially its present form. Stokes devised a method of inserting special cards in slots inside the bowl (now made of metal) of the Campbell instrument, one card being employed for each day's observations. This is the method still in use today.

About 1890 Whipple and Cassella, of London, started marketing the Campbell–Stokes recorder, the present design of which is shown in Fig. 8.3. The receiving surface is one of three types of specially designed pasteboard cards which is fitted into holding slots in the inside of the surrounding metallic spherical segment of the instrument. The type of card used is dependent on season; long curved cards are required in summer,

Fig. 8.3 A well-known model of the Campbell–Stokes sunshine recorder.

short curved cards in winter and straight cards in periods near the equin-
oxes. The color of the cards is normally a medium blue, and the composition is such that their dimensions do not expand appreciably on wetting.
A mechanism for moving the card carrier with respect to the glass sphere,
in order to adjust the position of the trace on the card, was incorporated
into an early design by Curtis (1898).

In 1962 the Commission for Instruments and Methods of Observation
of the World Meteorological Organization (WMO, 1965) adopted the
Campbell–Stokes sunshine recorder as a standard of reference known as
the "interim reference sunshine recorder" (IRSR) and recommended that
all future values of the duration of sunshine be reduced to the IRSR. It was
considered that by very careful determination of an instrumental reduction
factor, it should be possible to achieve international uniformity such that
systematic differences of measurement of the duration of sunshine would
not exceed $\pm$ 5%.

In order to use a Campbell–Stokes as an "interim reference sunshine
recorder," the instrument must be certified by the British Meteorological
Office, and some very careful adjustments are required. The main ones are
the following (WMO, 1969):

a.   The base must be leveled.
b.   The spherical segment must be adjusted so that the center line
     of the equinoctial card lies in the celestial equator. (The scale of
     latitude marked on the bowl support facilitates this.)
c.   The vertical plane through the center of the sphere and the noon
     mark on the spherical segment must be in the plane of the geo-
     graphic meridian.

The record cards also have strict specifications which apply when the
instrument is used as an IRSR. Thickness should be 0.4 ± 0.05 mm and
the width of the cards should be accurate to $\pm$ 0.3 mm. Changes in any
dimension with humidity should not exceed 2%. The color should correspond to a medium blue standard color, and the hour lines should be
black. Further detailed specifications for the cards have been issued by the
French Meteorological Office, and the cards should be certified by that
organization.

One problem in using the instrument is that it is difficult to define a
precise lower limit of direct radiant flux which will give a legible trace on
the card. In extreme conditions of a clear, dry atmosphere and a very
dry card, the threshold is as low as 7 mW cm^{-2}, while in the opposite
extreme conditions the threshold value is increased by as much as a factor
of 4. An average threshold value is probably about 21 mW cm^{-2} (WMO,
1965). Attempts have been made, with doubtful success, to interpret the

depth and breadth of the burn in terms of solar intensity. Detailed instructions for the use of the instrument are given by the British Meteorological Office (1956).

A special model of the Campbell–Stokes recorder has been developed for use at low-latitude stations, in which case the Sun attains such high angles that a pedestal mount for the glass sphere does not suffice. In this tropical model the sphere is held inside a semicircular brass sector, by two small cups mounted at the ends of the sector, and pressed against the sphere at opposite ends of a diameter by adjusting screws. The sector, which is clamped to the base of the instrument, may be adjusted for the latitude of the station. The record cards and methods of data retrieval are the same for the tropical model as for the standard model, as are the type of installation and adjustments which have to be made.

8.3.2 Maring–Marvin Sunshine Recorder

The Maring–Marvin sunshine recorder also operates from the heating effect of direct solar radiation, but in this instrument the actuating mechanism is the temperature difference induced between a blackened, highly absorbing glass bulb and a clear glass bulb which absorbs little of the solar energy. The deviation in temperature of the bulbs causes a differential expansion of a mixture of air and alcohol vapor in the bulbs, which, in turn results in movement of a mercury column and thereby closes an electric switch. The switch is normally in an appropriate electrical circuit for remote recording by means of some type of chronograph or other time-dependent mechanism.

Although the instrument has often been named for C. F. Marvin, chief of the U.S. Weather Bureau* from 1913 to 1934, the original concept was that of D. T. Maring (1898) in 1891. Marvin's design of a sunshine recorder in 1888 was a modification of the previous photographic instrument designed by J. B. Jordan in 1885 (see Section 8.2.2). Both types were apparently used simultaneously at the Weather Bureau stations in the 1890s. Maring's "thermometric" sunshine recorder passed through at least three modifications between 1891 and 1898, the instrument we now designate as the Maring–Marvin sunshine recorder perhaps most resembling the model of 1895, shown in Fig. 8.4. Maring's model of 1897 was an elaborate affair which combined the thermometric and photographic principles to record on a clock-driven photographic chart the height of the column of mercury. By this means, some knowledge of the temperature

---

* Now the U. S. National Weather Service (of the National Oceanic and Atmospheric Administration).

Fig. 8.4 Maring–Marvin sunshine recorder (1895 model) (courtesy of the National Oceanic and Atmospheric Administration).

difference between the clear and blackened bulbs, and thence a rough idea of the sunshine intensity, could be obtained. Although Maring obtained records with the instrument in the fall of 1897, it apparently was never put into general use.

A number of difficulties with interpretation of the records of the Maring–Marvin sunshine recorder are inherent in its design. A gross adjustment of the instrument involves a somewhat subjective procedure of shaking it in various positions (U.S. Weather Bureau, 1923) and the final adjustment is made by inclining the axis of the instrument in the meridian plane at such an angle that the mercury column will effect contact closure "during times when the disk of the Sun can be just faintly seen through the clouds." In an effort, not entirely successful, to eliminate effects of ambient temperature on the instrument, it is normally enclosed in an evacuated glass envelope. The residual temperature dependence, however, results in a recommendation that adjustments be made in winter or spring. Additional deficiencies are that the instrument does not respond to weak sunlight, it is affected by diffuse as well as direct radiation, and it is basically fragile. No exact standard for the instrument exists and, in general, a determination of sunshine duration with one such instrument does not entirely agree with that determined with another instrument of the same type or with the results from a Campbell–Stokes sunshine recorder. The main advantage of the Maring–Marvin sunshine recorder is the possibility it provides for remote recording. Although the Maring–Marvin type of device is still in use in some locations it has been largely replaced in the United States meteorological network by the Foster sunshine switch.

8.3.3   Foster Sunshine Switch

The Foster sunshine switch was developed by the U.S. Weather Bureau and reported on in 1953 (Foster and Foskett). The basic sensor is a pair of selenium barrier layer photovoltaic cells which are mounted so that one is shielded from the direct solar radiation by a shading ring (Fig. 8.5), while the other is exposed. Both cells "see" the diffuse skylight. They are connected in electrical opposition so that their responses to diffuse light result in no electrical signal. Direct sunlight, however, produces a signal from the exposed cell which is not balanced by an opposing signal from the shaded cell, the imbalance being used to trip a sensitive relay and actuate a recorder.

The two selenium cells are hermetically sealed and mounted in opposite ends of an opal-glass cylinder approximately 4.5 in. long and 2 in. in diameter. The cells are optically separated by a diaphragm across the cylinder at a point equidistant from the ends, and the middle section of the cylinder is covered by an opaque metal shield. The translucent character of the exposed ends of the cylinder produces internal light scattering which makes the sensitivity almost independent of the angle of incidence of the light. The slotted plate, which can be seen mounted above the glass cylinder in Fig. 8.5, protects the photocells from the full intensity of the Sun when it is high, but permits full exposure with maximum sensitivity during periods near sunrise and sunset.

In operation, the instrument is mounted with the axis of the cylinder parallel to the Earth's axis. The shading ring, which is 2 in. wide and 10 in. in diameter, can be set at various positions along the axis of the instrument, the appropriate position being a function of solar declination. It has been designed so that only four changes of position of the ring are required during the entire year.

Operation of the Foster sunshine switch over the period since its adoption has proved it to be stable and reliable, and very little maintenance is required. The lag of the instrument is negligible, and its sensitivity is sufficient to yield meaningful measurements throughout the sunrise to

Fig. 8.5 Foster sunshine switch (courtesy of the National Weather Service/NOAA).

sunset period. It is a standard instrument of the U. S. National Weather Service, and it has almost completely supplanted the Maring–Marvin sunshine recorder in the United States observational network.

8.4   INSTALLATION OF SUNSHINE RECORDERS

In a proper installation, the sunshine recorder must be illuminated by the direct solar beam throughout the time the Sun is above the horizon. Obstructions above the horizons in the east and west quadrants are to be particularly avoided. However, in view of the failure of most sunshine recorders to respond when the Sun is very near the horizon, obstructions up to about 3° above the horizon may be neglected for practical purposes. The permissible elevation angle for obstructions at other azimuths can be computed from the well-known formula for solar elevation expressed as

$$\sin \gamma = \sin \delta \sin \phi + \cos \delta \cos \phi \cos h \qquad (8.1)$$

where ϕ is latitude of the station, δ is solar declination and h is hour angle of the Sun. The calculation should be made for the time of the winter solstice ($\delta = \pm 23\frac{1}{2}°$), as the apparent path of the Sun is lowest on the horizon on that date. In case it is not possible to find a location completely free from obstructions, an estimate of the possible duration of sunshine which will be cut off during the different months should be obtained in order that corrections can be applied to the data. The estimate should be determined by the use of Eq. (8.1) and the measured outline of the the obstacle as viewed from the position of the sunshine recorder.

Details of the installation depend on the type of sunshine recorder to be installed. The Maring–Marvin instrument should be aligned with the meridian and inclined at such an angle that the mercury column will effect contact closure during times when the disk of the Sun can be just faintly seen through the clouds. The axis of the Foster sunshine switch should be aligned with the axis of the Earth. Probably the best method of effecting this alignment is to orient the axis toward the pole star by sight on a clear night. The shadow ring on the Foster sunshine switch is sufficiently wide so as to need adjustment only four times per year.

Alignment and adjustments are most critical for the Campbell–Stokes sunshine recorder. Detailed instructions for operation of the instrument and for measurement of the records are given by the Meteorological Office, London (1956). The main requirements which must be adhered to are the following:

1. The center of the sphere must be coincident with that of the bowl. This condition is met by the instrument manufacturer and adjustments should not be attempted in the field. Poor adjustments for concentricity cause the burns to be broad and ill defined at the edges.

2. The plane containing the central longitudinal line of the mounted equinoctial card must pass through the center of the sphere and coincide with the plane of the celestial equator. This condition is not met if there is an error in inclination of the instrument in the meridian plane or if it is tilted in a plane normal to the meridian. An error in inclination will cause a curvature of the burn at the time of the equinox and the burn will not be parallel to the edges of the cards in summer and winter. If the instrument is tilted in the normal plane, the burn is not symmetrical with respect to noon. If the burn is higher on the card in the morning than in the evening the west side of the instrument is higher than the east side, and vice versa.

3. The vertical plane through the center of the sphere which passes symmetrically through the bowl must coincide with the plane of the meridian. This condition is not met if the instrument is either tilted in the plane normal to the meridian (i.e., not level—condition 2 above) or if it is oriented incorrectly in azimuth. In either case, the burn will be asymmetric with respect to noon. Once the instrument is properly leveled, the azimuth adjustment can be made by positioning the instrument so that the solar

image coincides with the noon line on the card at the time of local apparent noon.

Measurement of the records of the Campbell–Stokes instrument is somewhat exacting, and a certain amount of judgment is involved. Instructions for the measurements are given in various publications, including Levert (1961) and particularly the Meteorological Office, London (1956). They are well summarized by the World Meteorological Organization (1965), which we quote as follows (pp. IX. 34–35):

(a) In the case of a clear burn with round ends, the length should be reduced at each end by an amount equal to half the radius of curvature of the end of the burn; this will normally correspond to a reduction of the overall length of each burn by 0.1 hour;

(b) In the case of circular burns, the length measured should be equal to half the diameter of the burn. If more than one circular burn occurs on the daily record it is sufficient to consider 2 or 3 burns as equivalent to 0.1 hour of sunshine; 4, 5, 6 burns as equivalent to 0.2 hour of sunshine; and so on in steps of 0.1 hour;

(c) Where the mark is only a narrow line, the whole length of this mark should be measured, even when the card is only slightly discolored;

(d) Where a clear burn is temporarily reduced in width by at least a third, an amount of 0.1 hour should be subtracted from the total length for each such reduction in width, but the maximum subtracted should not exceed one half of the total length of the burn.

8.5   CORRELATION BETWEEN SUNSHINE DURATION AND GLOBAL RADIATION

A number of studies for estimating the flux of global (direct plus diffuse) radiation from records of sunshine duration have been conducted, in an effort to supplement the measurements obtained from the radiation measuring stations of the observation network. The method, first suggested by Kimball (1919), was placed on a mathematical basis by A. K. Ångström (1924) about 5 years later. It has been used in its original form and with variations by many authors over the several decades since that time. A. K. Ångström derived the linear relation between the total daily global flux R and percentage of possible sunshine S for the latitude and date, which may be written in the form

$$R = R_c[a + (1 - a)S] \qquad (8.2)$$

where R_c is the flux for a perfectly cloudless day and a is an empirically determined constant. Because R_c varies with atmospheric transmission from day to day, many workers have followed Prescott (1940) in using the   relation

$$R = R_0(a - bS) \tag{8.3}$$

where R_0 is the solar flux on a horizontal surface at the top of the atmosphere, and a and b are constants to be determined from measurements. Harris (1966) tested two nonlinear regressions involving a third constant c in the form

$$R = R_0(a + bS + cS^2) \tag{8.4}$$

$$R = R_0(a + bS + c \log S) \tag{8.5}$$

together with the linear Eqs. (8.2) and (8.3); he found that the nonlinear relations gave slightly better results at very low values of R and S. This conclusion was reached also by McQuigg and Decker (1958). However, the low values of R generally contribute such a small part of the total flux, integrated over a period of a day, that the improvement by the nonlinear forms is relatively insignificant except at locations which have a high degree of cloudiness.

In an extensive study based on Eq. (8.3), Baker and Haines (1969) computed correlations between radiation fluxes measured at 4 radiation stations in Alaska and 15 stations in the north-central United States with those estimated from 3 to 6 sunshine duration stations either at or surrounding each radiation station. It was found that the correlations between observed and estimated fluxes were not uniformly high, even for data taken at the same station. The discrepancies were attributed to the following five factors:

1. The time near sunrise and sunset during which direct radiation is too weak to activate the sunshine recorder is a variable which depends on turbidity and water vapor content of the atmosphere.

2. Records of sunshine duration do not indicate radiant intensity, whereas that is the quantity measured by solar radiation instruments.

3. Paradoxical situations occur in which the highest radiation values are measured on partly cloudy days (with the Sun unobscured).

4. The two types of stations are often in different geographical locations, resulting in different atmospheric conditions for the measurements.

5. There may be undetected instrumental or observational errors.

McQuigg and Decker (1958) analyzed radiation and sunshine duration records for an 11-year period at Columbia, Missouri by use of the linear regression of Eq. (8.3); they obtained the results shown in Fig. 8.6 for the different seasons. It was found that from 62 to 91% of the variations in the

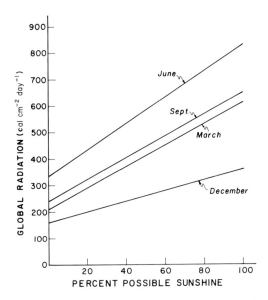

Fig. 8.6 Regression curves for estimating the global radiation from measurements of sunshine duration for different seasons at Columbia, Missouri (redrawn from McQuigg and Decker, 1958).

daily global flux of radiation could be explained on the basis of sunshine duration. By use of the same relation, Rosenberg (1964) found that at Rapid City, South Dakota and Dodge City, Kansas 80 and 90% respectively, of the flux variations can be explained by sunshine duration. At Lincoln, Nebraska, however, the value varies between 50 and 93%, the large range presumably arising from inconsistency in the atmospheric conditions at Lincoln.

Lof *et al.* (1966) made the logical assumption that the values of the constants of Eq. (8.3) are closely associated with the climatic regime, and used records from selected climatological stations to develop a world map of solar radiation for each month of the year. Their results of estimations of solar radiation from records of sunshine duration are summarized in Table 8.1. The climatic classification is that of Trewartha (1954). It can be seen, from the tabulation, that the values of a and b were found to vary by factors of about 3 and 8, respectively, depending on type of climate and location. It appears, however, that any clear patterns are obscured by the variations of the data.

The use of relative humidity as an additional parameter to improve estimates of radiation based on sunshine duration in the tropics was studied

TABLE 8.1

*Constants a and b and Average Percent Possible Sunshine S
for Various Locations and Climatic Types*[a]

Location	Type of climate	a	b	S
Miami, Florida	Tropical rainy, winter dry	0.42	0.22	65
Honolulu, Hawaii	Tropical rainy, no dry season	0.14	0.73	65
Stanleyville, Congo	Tropical rainy, no dry season	0.28	0.39	48
Poona, India	Tropical rainy, no dry season	0.30	0.51	37
Malange, Angola	Tropical rainy, winter dry	0.34	0.34	58
Ely, Nevada	Dry, arid (desert)	0.54	0.18	77
Tamanrasset, Sahara	Dry, arid (desert)	0.30	0.43	83
El Paso, Texas	Dry, arid (desert)	0.54	0.20	84
Albuquerque, New Mexico	Dry, arid (desert)	0.41	0.37	78
Brownsville, Texas	Dry, semiarid (steppe)	0.35	0.31	62
Charleston, South Carolina	Humid mesothermal, no dry season	0.48	0.09	67
Atlanta, Georgia	Humid mesothermal, no dry season	0.38	0.26	59
Buenos Aires, Argentina	Humid mesothermal, no dry season	0.26	0.50	59
Hamburg, Germany	Humid mesothermal, no dry season	0.22	0.57	36
Nice, France	Humid mesothermal, dry summer	0.17	0.63	61
Madison, Wisconsin	Humid continental, no dry season	0.30	0.34	58
Blue Hill, Massachusetts	Humid continental, no dry season	0.22	0.50	52
Dairen, Manchuria	Humid continental, dry winter	0.36	0.23	67

[a] After Lof *et al.*, 1966.

by Swartman and Ogunlade (1967) in the following three relations:

$$H = AS^a r^b \tag{8.6}$$

$$H = Be^{b(S-r)} \tag{8.7}$$

$$H = C + aS + br \tag{8.8}$$

Here r is relative humidity as measured at a height of about 5 feet above the ground, and A, B, C, a, and b are all constants to be determined from measurements. By using data from Nigeria, they obtained somewhat better correlations between solar radiation and sunshine duration than that given by Eq. (8.3), but the improvement was small and probably not significant.

In summarizing the above results, it may be concluded that the use of sunshine duration measurements to estimate the global flux of solar radiation is based on sound physical principles, and that the results are valuable in many locations to extend the relatively few records of global flux which may be available. However, for this purpose the method must be viewed as a poor second to global radiation measurement programs.

REFERENCES

Ångström, A. K. (1924). Solar and terrestrial radiation. *Quart. J. Roy. Meteorol. Soc.* **50**, 121–125.

Baker, D. C., and Haines, D. A. (1969). Solar radiation and sunshine duration relationships in the north-central region and Alaska. Tech. Bull. 262, Agr. Exp. Sta., Univ. of Minnesota. Minneapolis, Minnesota.

Bennett, I. (1967). Frequency of daily insolation in anglo North America during June and December. *Sol. Energy* **2**, 41–55.

Covert, R. N. (1925). Meteorological instruments and apparatus employed in the U.S. Weather Bureau. *J. Opt. Soc. Amer. Rev. Sci. Instrum.* **10**, 299–425.

Curtis, R. H. (1898). Sunshine recorders and their indications. *Quart. J. Roy. Meteorol. Soc.* **24**, 1–30.

Dines, W. H. (1900). The ether sunshine recorder. *Quart. J. Roy. Meteorol. Soc.* **26**, 243–246.

Foster, N. B., and Foskett, L. W. (1953). A photoelectric sunshine recorder. *Bull. Amer. Meteorol. Soc.* **34**, 212–215.

Glazebrook, Sir Richard. (1923). "A Dictionary of Applied Physics," Vol. 3, pp. 699–719. Macmillan, London.

Harris, A. R. (1966). Solar radiation reception and its correlation with sunshine. Masters thesis, Soil Sci. Dept., Univ. of Minnesota, Minneapolis, Minnesota.

Kimball, H. H. (1919). Variations in the total and luminous solar radiation with geographical position in the United States. *Mon. Weather. Rev.* **47**, 769–793.

Levert, C. (1961). Uber die auswertung der registrierungen des sonnenscheinautographen Campbell-Stokes. *Arch. Meteorol. Geophys. Bioklimatol., Ser. B* **2**, 135–137.

Lof, G. O. G., Duffie, J. A., and Smith, C. O. (1966). World distribution of solar radiation. *Sol. Energy* **10**, 27–37.

Maring, D. T. (1898). An improved sunshine recorder. U. S. Dept. Agr. Misc. Pub., Weather Bur. Circ. No. 148.

Marvin, C. F. (1896). Care and management of sunshine recorders. U. S. Weather Bureau, Circ. G, No. 109, p. 7.

McQuigg, J. D., and Decker, W. L. (1958). Solar energy—A summary of records at Columbia, Missouri. *Mon. Agr. Exp. Sta. Res. Bull.* 671.

Meteorological Office, London (1956). "Handbook of Meteorological Instruments: Part I—Instruments for Surface Observations." Her Majesty's Stationery Office, London.

Middleton, W. E. K. (1969). "Invention of the Meteorological Instruments." Johns Hopkins Press, Baltimore, Maryland.

Prescott, J. A. (1940). Evaporation from a water surface in relation to solar radiation. *Trans. Roy. Soc. So. Aust.*, **64**, 114–125.

Rosenberg, N. J. (1964). Solar energy and sunshine in Nebraska, Neb. Agr. Exp. Sta. Bull. No. 213.

Swartman, R. K., and Ogunlade, O. (1967). Solar radiation estimates from common parameters. *Sol. Energy* **2**, 170–172.

Trewartha, G. T. (1954). "An Introduction to Climate." McGraw–Hill, New York.

U. S. Weather Bureau (1923). Circ. G, Instrum. Div., 5th ed., Washington, D.C.

World Meteorological Organization (1965). Measurement of radiation and sunshine. Guide to Meteorological Instruments and Observing Practices, 2nd ed., W.M.O. 8, TP3 (Loose Leaf—1961).

CHAPTER NINE

The Solar Constant

The solar constant, defined as the flux of solar radiant energy across a surface of unit area oriented normal to the solar beam at the mean Sun–Earth distance, is an important physical quantity for many reasons. Solar radiation is the fundamental source of energy for driving the general circulations of the atmosphere and oceans; sunlight reaching the surface promotes photosynthesis in plants, without which life could not exist on Earth. Solar energy degrades spacecraft coatings and presents a hazard to astronauts in space. It is responsible for the ionosphere, which is so important in radio communications. In fact, solar radiation is involved in some way, either directly or indirectly, with most processes taking place in the Earth-atmosphere system.

The Sun itself is a gaseous sphere, the temperature of which varies from about 6000°K at the surface of the photosphere (the surface of the Sun as seen in white light) to over 1×10^6 °K in the corona (the tenuous outer atmosphere of the Sun), and over 1×10^7 °K in the deep interior.

This high interior temperature promotes the conversion of hydrogen into helium, which is the process responsible for the great amount of energy emitted by the sun. Each square centimeter of the Sun emits energy at the rate of 6.2 kW or 9.0×10^4 cal min^{-1}. Only about 1/2,000,000,000 of the energy emitted by the Sun reaches the earth, but this small fraction constitutes 1.60×10^{14} kW for the Earth as a whole (more than 500,000 times the rate at which electricity is generated in the United States).

TABLE 9.1

Physical Properties of the Sun and Earth[a]

Linear diameter of Sun	1.39196×10^{-6} km
Angular diameter of Sun at mean Earth–Sun distance	32 min, 2.4 sec
Mass of Sun	1.989×10^{33} g
Volume of Sun	1.4122×10^{33} cm³
Mean density of Sun	1.409 g cm⁻³
Mass ratio Sun: Earth	332,488
Surface gravity of Sun	2.7398×10^4 cm sec⁻²
Temperature of solar corona	$\sim 1,000,000°$K
Effective blackbody temperature of photosphere	5800°K
Approximate composition of Sun (by weight)	
Hydrogen	75%
Helium	24.25%
Heavy elements	0.75%
Equatorial radius of Earth	6378.17 km
Polar radius of Earth	6356.79 km
Mass of Earth	5.977×10^{27} g
Volume of Earth	1.08322×10^{27} cm³
Mean density of Earth	5.517 g cm⁻³
Surface gravity of Earth (standard)	980.665 cm sec⁻²
Mean Earth–Sun distance	1.495985×10^8 km
Value of solar constant	1.94 ± 0.02 cal cm⁻² min⁻¹
	1.353×10^6 erg cm⁻² sec⁻¹

[a] Principally after Allen, 1963.

Some of the physical dimensions related to the Sun, as given by Allen (1963), are listed in Table 9.1.

9.1 DETERMINATIONS OF THE SOLAR CONSTANT

9.1.1 Observations by the Smithsonian Institution

One of the great scientific investigations of history, and one which serves as a model of persistence and dedication to a single purpose, is the work of the Smithsonian Institution in the simultaneous development of pyrheliometers as scientific instruments and the establishment of the value of the solar constant of radiation. As early as 1873, Samuel Pierpont Langley (1834–1906) became interested in the Sun, making minute telescopic drawings of sunspots and studying infrared spectra during the period 1873–1880. About 1880 he invented the bolometer by means of which he was able to study the spectrum of solar radiation.

The first of the numerous scientific expeditions conducted by the Smithsonian Institution for solar radiation measurements was Langley's expedition to the top of Mt. Whitney in 1881. As befits a first, it was carried out with a flair, having a private railraod car for the cross-country leg of the trip and a mounted guard of Army cavalrymen from rails-end to the mountain. From an altitude of 12,000 feet on Mt. Whitney, Langley observed the solar infrared spectrum to extend beyond 1.8 μm, thereby overturning the previously held idea that the solar spectrum would end at about 1 μm. Langley called the long-wavelength region the "New Spectrum" (Langley, 1900). Before the end of the nineteenth century he had mapped out 740 absorption lines in the solar spectrum (Langley, 1902), and had obtained a very rough estimate of the solar constant as 3 cal cm^{-2} min^{-1}. At this time estimated values ranged from 1.75 to 4.0 cal cm^{-2} min^{-1} (Abbot *et al.*, 1913).

The solar radiation work of the Smithsonian Institution which had been started so brilliantly by Langley was picked up about the beginning of the twentieth century by Charles Greeley Abbot (1872–1973) and his colleagues, mainly F. E. Fowle, L. B. Aldrich, and W. H. Hoover. For more than 60 years, in spite of numerous local conflicts and two world wars, the investigation was pursued. When the regular observers were taken into the military, their wives operated the stations so as to prevent an interruption of the observations. While the longest-term stations were at Mount Montezuma, Chile (established in 1920, closed in 1955) and Table Mountain, California (established in 1925, closed about 1962), shorter-term stations were set up at Mount Wilson, on the Sinai Peninsula, in Algeria, and in other desert-type locations. Annual expeditions were made to the locations. Expeditions were conducted to the top of Mount Wilson from 1905 to 1912 and to the top of Mount Whitney in 1909 and 1910. In addition, a total of eight balloon flights were made shortly before World War I. Instrument development was carried along with the observations, and resulted in the silver-disk, water-flow, and water-stir pryheliometers and a special pyrheliometer for balloon application.

The main uncertainty inherent in the Smithsonian solar constant determinations, as well as in all other determinations made within the atmosphere, is that introduced by the atmosphere itself. Water vapor, ozone, dust, and other atmospheric gases and aerosols attenuate the incident solar beam by amounts which vary with atmospheric conditions, and the measurements are further contaminated by scattered light, particularly from the region of the sky near the Sun, entering the aperture of the instrument. Two different methods were developed in the Smithsonian program for taking account of atmospheric effects. In the "long method" several observations were made at different times, and consequently at different solar zenith angles, during a single day. As discussed in Chapter 3, the flux F_λ of direct radiation at wavelength λ which is transmitted by the atmosphere is related to the extraterrestrial flux $F_{\lambda 0}$ by the Bouguer–Lambert law written as

$$F_\lambda = F_{\lambda 0} e^{-\tau_\lambda m} \tag{9.1}$$

$$\ln F_\lambda = \ln F_{\lambda 0} - \tau_\lambda m \tag{9.2}$$

where m is air mass (approximately sec θ_0).

If the atmospheric properties do not change during the series of observations, then τ_λ is constant, and a plot of $\ln F_\lambda$ versus m results in a straight line which may be extrapolated to give $F_{\lambda 0}$ at $m = 0$ (top of the atmosphere). Integration of $F_{\lambda 0}$ over wavelength yields the total flux F_0 incident at the top of the atmosphere, and reduction to the mean Sun–Earth distance by the inverse square law yields a value of the solar constant.

There are the following four main sources of error inherent in the "long-method" determination.

1. An unknown amount of absorption by water vapor and carbon dioxide, principally in the near-infrared spectral region. This was taken into account by measuring the solar spectrum with a spectrobolometer and correcting for the absorption bands.

2. An unknown amount of ultraviolet absorption by ozone. Since all radiation of $\lambda < 0.2950$ μm is absorbed in the upper atmosphere, it was only after the advent of rockets that this part of the spectrum was measured.

3. An unknown amount of radiation from the bright circumsolar sky entering the aperture of the observing instrument.

4. Variations in atmospheric properties during a series of measurements.

In spite of very careful evaluations of each of these factors and a tremendous number of observations, a certain amount of error was unavoidable.

By the Smithsonian "long method," each determination required 2 to 3 hr of observation time, plus twice that much time for data reduction. Aside from the burdensome task of carrying out the work, there was no assurance that the atmospheric properties did not change in some consistent manner during the observational period. Because of these difficulties, a "short method" was devised by use of the data already obtained by the long method. If at a given observational site, the mean brightness of the sky and a mean amount of precipitable water had been determined, then changes from the means could be used to derive empirical relations for correcting a single observation for atmospheric effects. On this basis, one observation required only 10–15 min to complete, instead of the 7 to 8 hr necessary in the long method.

From many thousands of observations at several locations in different parts of the world and stretching over more than a half century, the best value of the solar constant as determined by the Smithsonian observations is 1.940 cal cm^{-2} min^{-1} or 139.5 mW cm^{-2} (Abbot, 1965). It is interesting that this is the same value which has recently been accepted as a standard for the solar constant.

9.1.2 Recent Solar Constant Determinations

A number of other determinations of the solar constant have been made, some of which have been modifications based on the data of the Smithsonian Institution, and some represent new and independent measurements from mountain tops, balloons, aircraft, or spacecraft. Johnson (1954) applied corrections to the Smithsonian observations in accord with new data on solar spectra in the ultraviolet and near-infrared regions, arriving at a value of 2.0 $\pm$ 2% cal cm^{-2} min^{-1}. Nicolet (1949) used a different method of spectral integration of the Smithsonian data in order to better account for Fraunhofer absorption, and obtained the value 1.981 $\pm$ 5%.

Other ground-based measurements have been made by Stair and Ellis (1968) and Labs and Neckel (1968). By the use of instrumentation from the National Bureau of Standards for observations at an altitude of 3660 m on Mauna Loa, Hawaii, Stair and Ellis combined new measurements in the 0.31–0.53 μm region with the previous data of Johnson (1954), obtaining a value of 1.95 cal cm^{-2} min^{-1} for the solar constant. In a remarkably complete discussion of the solar constant problem, Labs and Neckel (1968) evaluated the data of other observers and combined them with their own measurements taken from an altitude of 3600 m on the Jungfraujoch to arrive at a value of 1.958 cal cm^{-2} min^{-1}. Relatively high spectral irradiance values obtained from the observations of Makarova (1957) and Sitnik (1965) which were summarized by Makarova and Kharitonov (1969) and spectrally integrated by Thekaekara (1971) gave 2.03 cal cm^{-2} min^{-1}.

In order to minimize atmospheric effects in solar constant determinations, a number of measurements have been made from balloons floating in the 27–35-km altitude range, jet aircraft at about 12 km, the X-15 rocket aircraft at 82 km, and the Mars Mariner VI and VII spacecraft entirely outside the atmosphere. The advantage to be gained can be appreciated by realizing that even at 12 km a platform is above 79% of the atmospheric mass and above more than 99% of the water vapor. Only 1% of the atmosphere is above a balloon floating at an altitude of 31 km. The rocket aircraft and the Mariner spacecraft were effectively above the atmosphere; they were the only ones of the instrument platforms to reach altitudes above the ozone layer.

Six flights of the NASA aircraft Galileo were made at an altitude of about 12 km in 1967 (Thekaekara et al., 1968; Thekaekara, 1970). The aircraft was equipped with two Ångström and one Hy–Cal pyrheliometers and a cone radiometer for total flux measurements, as well as with two monochromators, two interferometers, and a multifilter radiometer for measurements of the solar spectrum. The results obtained in total flux measurements are shown in the first four entries of Table 9.2.

Balloon measurements of the solar constant were made at the University of Denver (Murcray et al., 1968; Murcray 1969) in three series of flights using Eppley pyrheliometers, and at the University of Leningrad (Kondratyev et al., 1967; Kondratyev and Nikolsky, 1970) at various times in the 1961–1966 period using Yanishevsky pyrheliometers. From

TABLE 9.2

Solar Constant Determinations from High-Altitude Measurements[a]

Platform (detector)	Solar constant ($cal\ cm^{-2}\ min^{-1}$)	Estimated error ($\pm\ cal\ cm^{-2}\ min^{-1}$)
Galileo aircraft (Hy–Cal pyrheliometer)	1.939	0.032
Galileo aircraft (Ångström 7635)	1.935	0.057
Galileo aircraft (Ångström 6618)	1.926	0.037
Galileo aircraft (Cone radiometer)	1.948	0.034
Balloon—Univ. of Leningrad (pyrheliometer)	1.919	0.009
Balloon—Univ. of Denver (pyrheliometer)	1.940	0.020
X-15 rocket aircraft (Special pyrheliometer)	1.951	0.033
Mariner spacecraft (cavity radiometer)	1.940	0.029

[a] The values for the two sets of balloon flights are reversed in Table II of Thekaekara's publication (after Thekaekara et al., 1968).

these series of measurements, Murcray derived a value of 1.940 and Kondratyev and his colleagues 1.919 cal cm^{-2} min^{-1}, with the estimated errors shown in Table 9.2.

The first direct measurement of the solar constant taken from above the ozone layer was made on 17 October 1967 from the X-15 rocket aircraft launched from Smith Ranch, Nevada (Laue and Drummond, 1968). The measurements were made in the 78–83-km altitude range, which is well above the ozone layer, with a special multichannel pyrheliometer built for the purpose (Drummond et al., 1967; Drummond and Hickey, 1968). The X-15 aircraft data yielded a value of 1.951 cal cm^{-2} min^{-1}.

Two series of measurements were made from completely above the atmosphere by means of special cavity radiometers mounted on the Mars Mariner VI and VII spacecraft over periods of several months while they were enroute to Mars. From the large number of measurements, which were corrected for changing distance from the Sun, a solar constant value of 1.940 cal cm^{-2} min^{-1} was obtained (Plamondon, 1969).

It is obvious from the range of values shown, which represent our best knowledge of the solar constant, that the last word has not yet been said on the subject. However, for spacecraft design and other engineering purposes, it is useful to have standard values of total solar radiation, as well as the spectral distribution of that radiation. After a thorough study of the available data, a special committee convened by NASA has arrived at a standard value of the solar constant and developed a standard of spectral irradiance to be used for design purposes (Thekaekara and

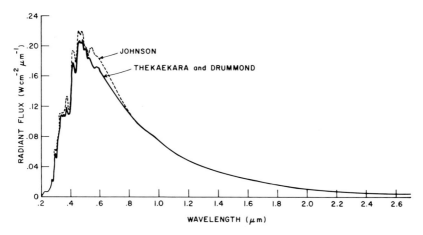

Fig. 9.1 The spectral distribution of the solar radiation incident on top of the atmosphere for a mean Sun–Earth distance (after Thekaekara and Drummond, 1971; Johnson, 1954).

Drummond, 1971). In developing the standard for the solar constant, the eight sets of measurements listed in Table 9.2 were summarized, and each set was weighted according to an estimate of its reliability. The final values were then taken as the weighted mean of the eight sets of measurements. Because of uncertainties of extrapolation to zero air mass, ground-based observations were not included in the evaluation. The standard of spectral irradiance was the result of a large number of spectral scans by two mono-chromators, two interferometers, and a filter radiometer, all of which were mounted on the Galileo aircraft.

As a result of this work, the standard value for the solar constant adopted by NASA is 1.940 ± 0.03 cal cm^{-2} min^{-1}, or 135.3 ± 2.1 mW cm^{-2}. This value is about 3% lower than the value obtained by Johnson (1954), which has heretofore been widely accepted. The standard of spectral irradiance, as given by Thekaekara and Drummond (1971), is tabulated in detail over the wavelength range 0.12–8.0 μm in Appendix C, and the range of 0.20–2.6 μm is plotted as the solid curve in Fig. 9.1. The main differences between this standard curve and that given by Johnson (1954) can be seen by comparing the curves in the diagram.

9.2   VARIABILITY OF SOLAR RADIATION

9.2.1   Annual Variation

Since the orbit of the earth is elliptical, the Sun–Earth distance varies throughout the year and causes a variation of the amount of solar energy reaching the Earth. Although the eccentricity of the orbit is small (only 0.01673), there is about 7% difference in the solar energy flux at the top of the atmosphere between perihelion and aphelion. The flux is greatest in early January and least in early July, as can be seen from the ratios of actual fluxes to their mean value listed in Table 9.3.

TABLE 9.3

Relative Fluxes F/F_0 of Solar Radiation Reaching the Top of the Atmosphere on the First Day of Each Month of the Year

Date	F/F_0	Date	F/F_0	Date	F/F_0
January 1	1.0335	May 1	0.9841	September 1	0.9828
February 1	1.0288	June 1	0.9714	October 1	0.9995
March 1	1.0173	July 1	0.9666	November 1	1.0164
April 1	1.0009	August 1	0.9709	December 1	1.0288

9.2.2   Other Variations of Solar Radiation

One question which is receiving increasingly more attention is the possible variability of the solar flux. If the solar constant is not really constant, then it is likely that the variations which exist will be accompanied by some phenomena which are important to man living on the Earth.

There are two principal means of detecting changes of the solar emission. The primary method is by direct observation by instruments measuring the emitted energy. Measurements by the Smithsonian Institution stretching over more than half a century are the main pool of data for this type of analysis, and they have been studied intensively. From such analysis, it has been well established that the early conclusion of Abbot *et al.* (1913) to the effect that the Sun's emission varies with an irregular periodicity of from 7 to 10 days with irregular amounts of the order of 10% is no longer tenable. Although the existence of such large and frequent variations has been discounted, the possibility of smaller variations is by no means ruled out. But it is difficult to obtain definitive evidence on the point. The Smithsonian and other measurements show that if such a variation of solar emission does exist, it is certainly not more than 1 or 2% of the solar constant, and probably considerably less than that. Thus it is difficult to separate out the actual variability from the unavoidable errors of measurement. For instance, from an analysis of the Smithsonian observations over the 1905–1917 time period, A. K. Ångström (1922) concluded that the solar constant varied by about 0.5%. However, as a result of later analyses of much more extensive Smithsonian observations, he (A. K. Ångström, 1970) concludes that changes of atmospheric transmission may account for all of the indicated variability of the solar constant. Other statistical studies of the data (Dorno, 1925; Clough, 1925; Page, 1939) tend to show that any actual variability is embedded in several types of errors, and is below the level of detection in the Smithsonian data. In fact, Clough (1925) concluded that changes of perhaps 0.1 to 0.2% are possible but stated ". . . one can scarcely fail to be impressed with the evidence for the almost absolute constancy of solar radiation."

The principal advocate of solar variability, but by no means the only one, is Dr. C. G. Abbot, who stoutly maintained at an age beyond his hundredth birthday, as he did over 40 years ago (Abbot, 1933), that ". . . major changes in the weather are due to short-period changes in the Sun." This position has been reiterated in numerous publications by Abbot (e.g., 1944, 1960, 1961, 1965), and has been supported by Clayton (1925), Aldrich (1945), Aldrich and Hoover (1954), Roberts (1952), and others. The evidence for solar constant variability has been bolstered more recently by independent measurements made by high-altitude balloon

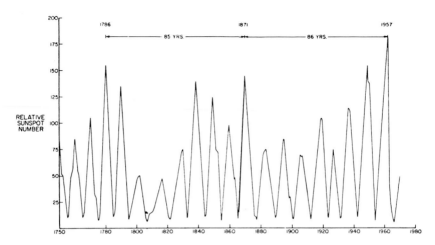

Fig. 9.2 The 11-year cycles of solar activity as recorded in Zurich, Switzerland since 1750 (after Newman, 1971).

by Kondratyev and Nikolsky (1970). These data will be mentioned further below.

As is well known, the Sun undergoes periodic changes of sunspot activity, the most obvious of such periods being the approximately 11-year cycle. An observatory in Zurich, Switzerland has been recording sunspot activity in terms of the Wolf Relative Sunspot Number for over 200 years. The data for that period, which are plotted in Fig. 9.2, show very clearly the periods and magnitudes of the solar cycles which have been recorded. We are now in cycle number 20. A somewhat more fundamental cycle consists of two of the 11-year cycles, at which 22-year period the magnetic fields associated with sunspots change polarity in the two hemispheres of the Sun. These periods are only approximate, as the actual periods vary somewhat in the different cycles.

Long- and short-period cycles of solar activity have also been reported by various authors. Willett (1965) has described an 80-year cycle of sunspot activity, and Simpson (1959) has suggested, on the basis of climatic change, that the solar constant undergoes cyclic variations with a period of 380,000 years. This latter value is based on controversial evidence and must be viewed with some skepticism until additional confirmation is obtained.

9.2.3 Short-Period Variations

The principal proponent of short-period solar cycles is Abbot (1933, 1944, 1956), who claims to have detected in the solar constant measurements a whole family of solar constant cycles consisting of at least 64 members, all of which are submultiples of 273 months (22.75 years) to within 1%. Over 200 of the periods have variations of 0.05% to 0.21% of the solar constant, according to Abbot. He made 60-year forecasts of temperature and precipitation on the basis of these short-period cycles, and found evidence from the weather records of St. Louis that a variation of 0.05 to 0.21% in the solar constant produces changes of 5 to 25% in the amount of precipitation. He suggests that solar variations are a leading weather element which should be observed on a regular basis.

Abbot's ideas have not been generally accepted among meteorologists or solar physicists. As mentioned above, however, independent direct evidence of variations of the solar constant with the sunspot cycle has been obtained from balloon observations by Kondratyev and Nikolsky (1970). Their data are consistent with the relation, first suggested by Ångström (1922), that the solar flux F_0 at the top of the atmosphere is given by

$$F_0 = 1.903 + k\sqrt{N} - cN^\alpha \tag{9.3}$$

where N is relative sunspot number and the constants have the values $k = 0.011$, $c = 0.0006$, and $\alpha \approx 1$.

It is likely that a real confirmation of the extent and magnitude of variations of the solar constant will have to await long-term measurements from satellites. Steps in this direction have already been taken, and additional experiments (particularly the Earth Radiation Budget experiment) are under development.

With respect to solar variations, we should mention the well-known fluctuations of the ultraviolet emission from the Sun. Even though the total amount of energy involved in ultraviolet variations is an extremely small fraction (perhaps 0.001%) of the solar constant, they are known to produce large effects in the upper atmosphere. Willett (1965) has developed his theory of climatic change on evidence of a modification of atmospheric circulation by the absorption of ultraviolet radiation.

The most definitive data on ultraviolet variations have been obtained from satellite observations by Heath (1969). In that experiment, radiation in three relatively broad spectral intervals in the 0.115–0.30 μm region was monitored during three 27-day solar rotations. Peak-to-peak variations were 7–37% of the constant background at 0.12 μm, about 6% at 0.18 μm, and only 0.9% at 0.26 μm. A solar flare increased the 0.12 μm flux by 16%, but had little effect at the other wavelengths.

In contradistinction to the direct measurement of solar constant variations, an alternate method is to observe cyclic changes of weather on the Earth and correlate them with known changes of sunspot activity. The inference, of course, is that if the cycles bear a definite relation to each other, then there is a cause and effect relationship between the physical phenomena. Many climatologists have taken this approach, and a plethora of conclusions have been drawn. Since the evidence is purely circumstantial, it is difficult to evaluate the validity of the conclusions.

A correlation of droughts in continental areas with the 11-year sunspot cycle has been summarized by Newman (1971). From a plot such as that of Fig. 9.2 it can be seen that sunspot cycles appear to alternate between major and minor maxima. For instance, the maximum at about 1928 was minor, the following one at about 1937 was major, that in 1948 was minor, and the one in 1957–1959 was major. According to Newman (1971), the minor maxima tend to be followed by dry periods, whereas no such effect is associated with major maxima. The 1928 minor maximum was followed by the dust-bowl days in the Midwest, and the early 1950's, following the 1948 minor maximum, was another dry period. By this reasoning, the middle 1970's should be a dry period, but superimposed on this 11-year cycle are cycles of 35–40 years and 80–90 years, which themselves have effects on the weather. These longer cycles would tend to make the 1970's wetter than usual, which would tend to counteract the drought normally expected from the 11-year cycle.

There is obviously no end to the possibilities in combinations of cycles, particularly if one accepts the 64 short-period cycles of Abbot. There does appear, however, to be some relatively convincing evidence of the existence of solar-related weather cycles in some localities. Intriguing correlations with sunspot activity have been found in the water level of rivers and lakes, the advance and retreat of glaciers, the general circulation of the atmosphere, temperature, pressure, humidity, and precipitation at various locations, and even in the occurrence of epidemics of malaria, whooping cough, diphtheria, and other diseases (Willett, 1965; Loginov, 1970; Loginov and Sazonov, 1971; Borotinskaya and Beliazo, 1969; Wagner, 1971; Konovalenko, 1970; Abbot, 1956; and many others). Unfortunately, the cyclic variations of weather elements are difficult to detect among the natural random fluctuations, as well as among errors of observation, and the statistical techniques employed have not always been adequate to the task. Thus caution must be exercised in accepting the reported results, but if properly applied this correlational technique seems promising as a useful tool in long-range weather forecasting.

REFERENCES

Abbot, C. G. (1933). Weather dominated by solar changes. *Smithson. Misc. Collect.* **85**, No. 1.

Abbot, C. G. (1944). Weather predetermined by solar variation. *Smithson. Misc. Collect.* **104**, No. 5.

Abbot, C. G. (1956). Periodic solar variation. *Smithson. Misc. Collect.* **128**, No. 4.

Abbot, C. G. (1960). A long–range forecast of United States precipitation. *Smithson. Misc. Collect.* **139**, No. 9.

Abbot, C. G. (1961). A long–range temperature forecast. *Smithson. Misc. Collect.* **139**, No. 5.

Abbot, C. G. (1965). The solar constant. *Solar Energy* **9**, 166–167.

Abbott, C. G., Fowle, F. E., and Aldrich, L. B. (1913). *Ann. Smithson. Astrophys. Obs.* **3**.

Aldrich, L. B. (1945). The solar constant and sunspot numbers. *Smithson. Misc. Collect.* **104**, No. 12.

Aldrich, L. B., and Hoover, W. H. (1954). *Ann. Smithson. Astrophys. Obs.* **7**.

Allen, C. W. (1963). "Astrophysical Quantities," 2nd ed. Oxford Univ. Press (Athlone), London and New York.

Ångström, A. K. (1922). The solar constant; the solar spots and solar activity. *Astrophys. J.* **55**, 24–29.

Ångström, A. K. (1970). Apparent solar constant variations and their relation to the variability of atmospheric transmission. *Tellus,* **22**, 205–218.

Borotinskaya, M. Sh., and Beliazo, V. A. (1969). Effect of solar activity on the formation of circulation epochs and their stages. *Tr., Arkt. Antarkt. Nauch. Issled. Inst.* **289**, 132–151.

Clayton, H. H. (1925). Solar variations. *Mon. Weather Rev.* **53**, 522–525.

Clough, H. W. (1925). A statistical analysis of solar radiation data. *Mon. Weather. Rev.* **53**, 343–348.

Dorno, C. (1925). Fluctuations in the values of the solar constant. *Mon. Weather. Rev.* **51**, 71–81.

Drummond, A. J., and Hickey, J. R. (1968). The Eppley-JPL solar constant measurement program. *Sol. Energy* **12**, 217–232.

Drummond, A. J., Hickey, J. R., Scholes, W. J., and Laue, E. G. (1967). Multi-channel radiometer measurement of solar irradiance. *J. Spacecr. Rockets,* **4**, 1200–1206.

Heath, D. F. (1969). Observations on the intensity and variability of the near ultraviolet solar flux from the Nimbus III satellite. *J. Atmos. Sci.* **26**, 1157–1160.

Johnson, F. S. (1954). The solar constant. *J. Meteorol.* **11**, 431–439.

Kondratyev, K. Ya., and Nikolsky, G. A. (1970). Solar radiation and solar activity. *Quart. J. Roy. Meteorol. Soc.* **96**, 509–522.

Kondratyev, K. Ya., Nikolsky, G. A., and Badinov, J. Ya. (1967). Direct solar radiation up to 30 km and stratification of attenuation components in the stratosphere. *Appl. Optics* **6**, 194–207.

Konovalenko, Z. P. (1970). Probable role of helio–geophysical factors in long–term variations of epidemic levels. *Geogr. Obs. SSSR, Izv.* **102**, 347–355.

Labs, D., and Neckel, H. (1968). The radiation of the solar photosphere from 2000 Å to 100 μ. *Zeit. Astrophys.* **69**, 1–73.

Langley, S. P. (1900). The new spectrum. *Smithson. Inst. Annu. Rep.* pp. 683–692.

Langley, S. P. (1902). The absorption lines in the infrared spectrum of the Sun. Smithson. Inst., Astrophys. Obs. Rep., Part 1, (1891–1901), pp. 7–21.

Laue, E. G., and Drummond, A. J. (1968). Solar constant: First direct measurements. *Science* **161**, 888–891.

Loginov, V. F. (1970). Oscillations of moisture in Europe and their possible cause. *Uch. Zap. Leningrad Gos. Univ.* **342**, 38–46.

Loginov, V. F., and Sazonov, B. I. (1971). Solar constant and the climate of the Earth. *Geogr. Obs. SSSR, Izv.* **103**, 229–233.

Makarova, E. A. (1957). A photometric investigation of the energy distribution in the continuous solar spectrum in absolute units. *Sov. Astron. AJ*, **1**, 531–546.

Makarova, E. A., and Kharitonov, A. V. (1969). Mean absolute energy distribution in the solar constant. *Sov. Astron. AJ*, **12**, 599–609.

Murcray, D. G. (1969). Balloon borne measurements of the solar constant. Rep. No. AFCRL 69-0070, Univ. Denver. Denver, Colorado.

Murcray, D. G., Kyle, T. G., Kosters, J. J., and Gast, P. R. (1968). The measurement of the solar constant from high altitude balloons. Rep. No. AFCRL 68-0452, Univ. Denver. Denver, Colorado.

Newman, J. E. (1971). Climatic changes: some evidence and implications. *Weatherwise* **24**, 56–62.

Nicolet, M. (1949). Sur le probleme de la constante solaire. *Ann. Astrophys.* **14**, 249–265.

Page, L. F. (1939). Comparison of contemporaneous measurements of the solar constant. *Mon Weather. Rev. Suppl.* **39**, 118–120.

Plamondon, J. A. (1969). The Mariner Mars 1969 temperature control flux monitor. Jet Propulsion Laboratory Space Programs Summary 37–59, **3**, 162–168, Pasadena, California.

Roberts, W. O. (1952). Stormy weather on the Sun. *Smithson. Inst. Annu. Rep.* pp. 163–174.

Simpson, G. C. (1959). World temperatures during the pleistocene. *Quart. J. Roy. Meteorol. Soc.* **85**, 332–349.

Sitnik, G. F. (1965). Results of two series of absolute photoelectric measurements of the solar spectrum. *Sov. Astron. AJ* **9**, 44–49.

Stair, R., and Ellis, H. T. (1968). The solar constant based on new spectral irradiance data from 3100 to 5300 Ångströms. *J. Ap. Meteorol.* **7**, 635–644.

Thekaekara, M. P. (1970). The solar constant and the solar spectrum measured from a research aircraft. NASA Tech. Rep. R-351, National Aeronautics and Space Administration. Washington, D. C.

Thekaekara, M. P. (1971). Solar electromagnetic radiation. NASA Space Vehicle Design Criteria (Environment), NASA SP-8005, National Aeronautics and Space Administration. Washington, D.C.

Thekaekara, M. P., and Drummond, A. J. (1971). Standard values of the solar constant and its spectral components. *Nature (London)* **229**, 6–9.

Thekaekara, M. P., Kruger, R., and Duncan, C. H. (1968). Solar irradiance measurements from a research aircraft. *Appl. Optics* **8**, 1713–1732.

Wagner, A. J. (1971). Long–period variations in seasonal sea–level pressure over the northern hemisphere. *Mon. Weather. Rev.* **99**, 49–66.

Willett, H. C. (1965). Solar–climatic relationships in the light of standardized climatic data. *J. Atmos. Sci.* **22**, 120–136.

CHAPTER TEN

Terrestrial Radiation: Field Characteristics

10.1 GENERAL

As pointed out previously, it is convenient for practical purposes to divide the entire atmospheric radiation regime into two parts—the solar (or short-wave) regime and the terrestrial (or long-wave) regime. This is made feasible because of the greatly differing temperatures of the Earth and Sun. The high temperature of the Sun (about 6000°K) results in over 99% of the solar energy being at wavelengths of less than 4 μm, whereas the much lower temperatures of the atmosphere and surface materials (generally <300°K) yields most energy in roughly the 4–100 μm region. Thus for purposes of overall energy considerations, a division of the spectrum at about 4 μm effectively separates the two.

The regime of terrestrial or thermal radiation of the atmosphere–surface system is in some respects more complicated than that of the solar radiation regime, although, of course, the same concepts and relationships are applicable to both. The greater complexity of the terrestrial regime is largely a result of the emission and absorption of such long-wave radiation by all real materials—solids, liquids, and gases—which constitute the physical system of interest. Soils, sands, vegetation, water, and other natural surfaces have emission and absorption characteristics approximating those of gray bodies (absorptivity A independent of wavelength but $A < 1.0$) and in some cases blackbodies ($A = 1.0$). By Kirchhoff's law, a good absorber is also a good emitter. Metals in general, and particularly polished metals, have low values of absorptivity (and emissivity) for long-wave radiation, as they do for short wavelengths. The radiative processes of most interest in the solar regime are absorption, reflection, and scattering by the constituents of the system, with emission of short-wavelength radiation being entirely negligible by comparison. In the terrestrial regime, on the other hand, emission is a major factor, radiation being emitted by the solid and liquid materials of the Earth's surface, by dust and clouds in the atmosphere, and by many atmospheric gases. Absorption by the same materials is also of prime concern in the terrestrial range, but the scattering of long-wave radiation plays a much less important role in the terrestrial than in the solar regime. Little is known about the polarization of terrestrial radiation in the atmosphere. At the present time no very satisfactory polarization measurement techniques are available for the longer wavelength range, and little work has been done in the area. It is likely, however, that future research will reveal some important polarization phenomena in the terrestrial range, although their nature is difficult to foresee.

The spectral distribution of radiation emitted by a blackbody source at various temperatures in the terrestrial range is shown in Fig. 10.1. The curves were obtained from a solution of Planck's law, given as Eq. (1.15) in Chapter 1. The maximum occurs at wavelengths of 10–12 μm, and the shift of the position of the maximum with temperature, as predicted by the Wein displacement law [Eq. (1.24)] is clearly evident in the diagram. A temperature of 300°K is typical for that of the Earth's surface and the lower

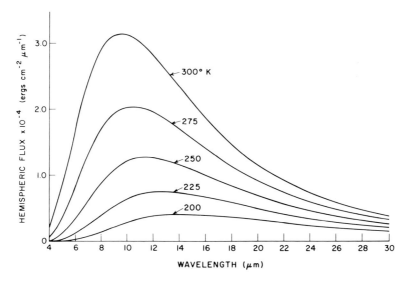

Fig. 10.1 Spectral distribution of radiation emitted by a blackbody at various temperatures which normally occur in the atmosphere.

levels of the atmosphere at low and midlatitude locations. Surface temperatures in the Arctic in winter are in the 250°K range, and stratospheric temperatures are normally in the 190–220°K range. It should be emphasized, however, that the curves are for blackbody radiation. The absorptivity (emissivity) of gases is strongly wavelength dependent, varying from low values in the "windows" to high values in the centers of strong absorption bands. Only in the wavelength regions of the strong bands to gases approximate blackbody radiators.

In general, a gaseous molecule can have three types of energy, in addition to the simple translational energy. The three additional types are electronic, vibrational, and rotational energies. Changes of these energies occur by absorption or emission of a discrete amount of energy (a quantum) which corresponds to a specific wavelength of radiation (a spectral line). Quanta for electronic transitions are at relatively high-energy levels (several electron volts) and their spectral lines are therefore confined to the visible and ultraviolet parts of the spectrum. Changes of rotational energy levels, on the other hand, require relatively little energy ($\sim 10^{-4}$ eV), and lines due to pure rotation are in the far-infrared and microwave regions of the spectrum. Energy levels of molecular vibrations are between the two ($\sim 10^{-1}$ eV). Since vibrational energies are roughly a thousand times greater than rotational energies, vibrational lines rarely occur alone, but usually in conjunction with rotational lines. The combination of molecular vibrations with rotations is responsible for groups of lines which form the many vibration-rotation bands in the near- and intermediate-infrared spectra of the atmospheric gases.

Gaseous absorption (and emission) of radiation in the atmosphere is mainly due to water vapor, carbon dioxide, ozone, and some minor constituents such as carbon monoxide, nitrous oxide, methane, and nitric oxide. The transmission characteristics of the Earth's atmosphere as a function of wavelength are shown at low spectral resolution in the upper part of Fig. 10.2. The gaseous absorbers responsible for the strongest bands are indicated in the diagram. The spectral distribution of energy emitted by the atmosphere with typical quantities of the radiatively active gases and an effective temperature of 273°K is shown in the lower half of the figure. Relatively little energy is emitted by the atmosphere in the "window" region between 8 and 13 μm, whereas in the strong bands of water vapor (5–8 μm and beyond 20 μm) and carbon dioxide (14–16 μm), atmospheric emission approximates that of a blackbody. The ground surface, on the other hand, emits approximately as a blackbody at all wavelengths. Thus terrestrial radiation emitted to space from the ground and clear atmosphere system is composed mainly of that emitted by the atmosphere in wavelengths of the strong absorption bands plus that which is emitted by the

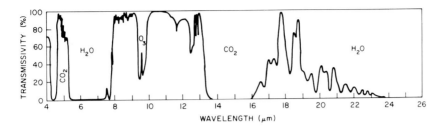

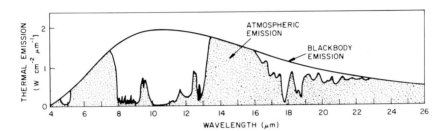

Fig. 10.2 (Upper) Spectral transmissivity of the Earth's atmosphere at wavelengths in the terrestrial radiation regime. The gases responsible for the main atmospheric absorption bands are indicated. (Lower) Spectral distribution of radiation emitted by the atmosphere compared to that emitted by a blackbody at the same temperature.

surface and transmitted outward in the regions of weak atmospheric absorption. The terrestrial radiation which reaches the ground from the clear sky above is mainly due to emission by atmospheric gases. The relatively dense liquid water clouds in the atmosphere act essentially as blackbodies, and thus strongly modify the fields of both upward and downward radiation.

10.2 EQUATION OF RADIATIVE TRANSFER

In analogy with the discussion of Chapter 1 (Section 1.1.3), we can write the equation of transfer for monochromatic radiation in a plane-parallel atmosphere which exhibits both emission and attenuation in the form

$$-\mu \frac{dI(\lambda; \tau; \mu)}{d\tau} = I(\lambda; \tau; \mu) - J(\lambda; \tau) \qquad (10.1)$$

Here $\mu = \cos \theta$, θ being the zenith angle, and τ is optical thickness of the medium. The first term on the right-hand side is responsible for attenuation in the medium and the second is responsible for emission by the medium itself. Thus the quantity $J(\lambda; \tau)$ is usually termed the source function. The source function is assumed to be isotropic and the intensity distribution in the plane-parallel atmosphere is taken to be independent of azimuth, both of which are close to reality under the usual distribution of radiators in the atmosphere. Equation (10.1) is known as the Schwarzschild equation of radiative transfer.

By definition, we have the relations

$$d\tau = k \, du \qquad (10.2)$$

and

$$J(\lambda; \tau) = e_\lambda/k_\lambda \qquad (10.3)$$

where u is optical mass of the absorbing gas, and k_λ and e_λ are the monochromatic absorption and emission coefficients, respectively. For a gas in local thermodynamic equilibrium (the case for the atmosphere below an altitude of roughly 50 km), the source function is equal to the Planck blackbody function.

The formal solution of Eq. (10.1) is given in any text on radiative transfer (e.g., Goody, 1964) and may be written

$$I(\lambda; \tau; \mu) = I(\lambda; \tau_1; \mu)e^{-(\tau_1-\tau)/\mu} + \int_\tau^{\tau_1} J(\lambda; t)e^{-(t-\tau)/\mu} \frac{dt}{\mu} \qquad (10.4)$$

Here τ_1, is the value of τ at the boundary and the variable of integration t varies from τ_1, to some reference value τ. We have adhered to the convention that τ increases in the downward direction.

The geometry of the problem may be seen from Fig. 10.3. The case

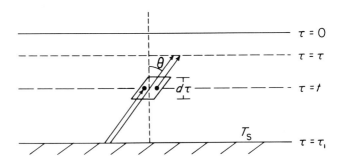

Fig. 10.3 Geometry of the radiative transfer problem for a plane-parallel atmosphere of normal optical thickness $\tau = \tau_1$ superimposed on an underlying surface of temperature T_s.

for the whole atmosphere is shown here, although the concepts apply to any atmospheric layer, and the direction of propagation is arbitrary. For convenience, we have shown radiation directed into the upward hemisphere at zenith angle θ (the $+\mu$ direction by convention). The surface indicated at $\tau = \tau_1$ is assumed to be the Earth's land or water surface, although it might also be the top of a cloud. It is usually assumed that the surface emission is independent of μ, although the measurements are not sufficient to establish the validity of this assumption, particularly for water surfaces and for clouds, and it is convenient to assume that $I(\lambda; \tau_1)$ is given by the black-body source function $B(\lambda; T_s)$ at a temperature T_s of the surface. Radiation incident from below at an arbitrary level $\tau = \tau$ is the sum of surface-emitted radiation which has been transmitted directly through the layer below $\tau = \tau$, being attenuated in the process, plus radiation which has been emitted at level t (which takes all possible values between τ_1 and τ) and likewise attenuated between t and τ. These two parts are represented by the first and second terms, respectively, on the right side of Eq. (10.4).

For radiation in the downward direction the same concepts apply, but in that case the radiation from the boundary ($\tau = \tau_0$: top of the atmosphere) is zero. Thus the first term on the right-hand side of Eq. (10.4) vanishes for downward radiation.

With the above assumptions, we can rewrite Eq. (10.4) for the intensity of radiation directed upward at angle $\theta(\mu > 0)$ through level τ in the form

$$I(\lambda; \tau; +\mu) = B(\lambda; T_s)e^{-(\tau_1-\tau)/\mu} + \int_\tau^{\tau_1} J(\lambda; t)e^{-(t-\tau)/\mu}\,\frac{dt}{\mu} \qquad (10.5)$$

and for downward radiation ($\mu < 0$) at level τ as

$$I(\lambda; \tau; -\mu) = \int_0^\tau J(\lambda; t)e^{-(\tau-\tau)/\mu}\,\frac{dt}{\mu} \qquad (10.6)$$

The total flux of radiant energy in the upward and downward directions across a horizontal surface at level τ is obtained by integrating Eqs. (10.5) and (10.6) over azimuth and elevation angles and over wavelength. Thus we have the relations

$$F(+\mu) = \int_0^\infty \int_0^{2\pi} \int_0^1 I(\lambda; \tau; +\mu)\,\mu\,d\mu\,d\phi\,d\lambda \qquad (10.7)$$

$$F(-\mu) = \int_0^\infty \int_0^{2\pi} \int_{-1}^0 I(\lambda; \tau; -\mu)\,\mu\,d\mu\,d\phi\,d\lambda \qquad (10.8)$$

10.3   RADIATIVE HEATING AND COOLING

In problems of radiative heating and cooling of the atmosphere, it is not the flux itself but the flux divergence which is the relevant quantity. The flux divergence at level τ is

$$\frac{dF}{d\tau} = 2\pi \int_0^\infty \int_{-1}^1 \frac{dI}{d\tau} \mu \, d\mu \, d\lambda \tag{10.9}$$

Here we have used the azimuth independence to obtain the factor 2π from the integration over azimuth, and the integration over μ takes account of both upward and downward radiation. The use of Eq. (10.2) permits us to write the flux divergence in terms of optical mass u as

$$\frac{dF}{du} = 2\pi \int_0^\infty \int_{-1}^1 \frac{dI}{du} \mu \, d\mu \, d\lambda \tag{10.10}$$

or, in terms of pressure,

$$\frac{dF}{dp} = 2\pi \int_0^\infty \int_{-1}^1 \frac{dI}{dp} \mu \, d\mu \, d\lambda \tag{10.11}$$

Finally, by introducing Eq. (10.1) we have

$$\frac{dF}{dp} = 2\pi \frac{du}{dp} \int_0^\infty \int_{-1}^1 [I(\lambda; \tau; \mu) - J(\lambda, \tau)] k_\lambda \, d\mu \, d\lambda \tag{10.12}$$

The rate of heating or cooling at level τ is simply

$$\frac{dT}{dt} = \frac{g}{C_p} \frac{dF}{dp} \tag{10.13}$$

where C_p is the specific heat at constant pressure and g is the acceleration of gravity. Thus, in principle at least, if we know the vertical profiles of the temperature and the radiatively active gases, we should be able to compute the rate of temperature change at any point in the atmosphere.

For practical purposes it is often important to determine the rate of heating or cooling averaged over a layer in the atmosphere, instead of that at one level. The average rate of change of temperature between pressure altitudes 1 and 2 is

$$\frac{dT}{dt} = \frac{g}{C_p} \left\{ \frac{[F(+\mu) - F(-\mu)]_2 - [F(+\mu) - F(-\mu)]_1}{|\Delta p|} \right\} \tag{10.14}$$

The determinations of the fluxes are, of course, subject to the difficulties mentioned above, plus the fact that the calculations, being based on small differences of large quantities, are very sensitive to computational errors. In spite of these difficulties, however, the methods are frequently used and are of sufficient accuracy for many practical problems.

10.4 MODELS OF ABSORPTION LINES AND BANDS

We turn now to the problem of determining reasonable approximations by which the equation of radiative transfer can be solved in the face of the complex line and band spectra exhibited by the atmospheric gases. We shall follow convention by using frequency ν instead of wavelength for the purpose. The fractional absorption $(A_\nu)_l$ of a single line can be expressed as

$$(A_\nu)_l = 1 - e^{-K_\nu u} \qquad (10.15)$$

where K_ν is the mass absorption coefficient at the frequency of the line, and u is optical mass. The absorption coefficient of a band is obtained by integration of (10.4) over the frequency interval of the band. Thus

$$A_\nu = \int_{\nu_1}^{\nu_2} (A_\nu)_l \, d\nu = \int_{\nu_1}^{\nu_2} \left[1 - e^{-K_\nu u}\right] d\nu \qquad (10.16)$$

The absorption coefficient depends on both line intensity and line shape. Line intensity, for the lines of most importance in the atmosphere, is a slowly varying function of temperature, and line shape is somewhat dependent on temperature but much more strongly on pressure.

For the pressure range in which the linewidth is determined primarily by collisions with other molecules (altitudes below about 50 km in the Earth's atmosphere), the actual line shapes are reasonably well approximated by the relation originally obtained by H. A. Lorentz in 1906 written as

$$K_\nu = Sb = S \left\{ \frac{\alpha}{\pi \left[(\nu - \nu_0)^2 + \alpha^2 \right]} \right\} \qquad (10.17)$$

where S is line intensity, α is the half-width of the line at one-half of the maximum of K_ν, ν_0 is frequency of the line center, and b is the line-shape factor normalized to unity by $\int_0^\infty b(\nu) \, d\nu = 1$. For pressures at which line broadening is due mainly to molecular velocities (altitudes above about 50 km in the atmosphere) the Doppler line shape applies, yielding

$$K_\nu = \frac{S}{\alpha_D} \left(\frac{\ln 2}{\pi} \right)^{1/2} \exp\{-[(\nu - \nu_0)/\alpha_D]^2 \ln 2\} \qquad (10.18)$$

where α_D is the Doppler width of the line. A Doppler broadened line is considerably narrower and has less absorption in the wings than does a Lorentz line.

In theory, by knowing the location, strength, and shape of the absorption lines, one could solve the equation of radiative transfer [Eqs. (10.4) and (10.5)] directly for an absorption band by performing a line-by-line calculation and integrating the results over the frequency of the band. This procedure has been used by Hitchfield and Houghton (1961) for a portion of the 9.6-μm band of ozone and by several authors for parts of the 15-μm carbon dioxide band. There are several problems with the direct approach, however. Not only is a great deal of computer time required for the thousands of lines constituting an absorption band, but a more fundamental difficulty is the lack of precise information on the positions, shapes, and strengths of the lines. If these data are not available for the lines for the simple linear molecule of carbon dioxide, they are much less well known for the complicated water vapor and ozone molecules. In addition, there is no simple means of adequately accounting for the overlap of the wings of neighboring lines. In spite of these difficulties, however, the direct method is useful for evaluating the validity of the more empirical methods discussed below.

The only practical alternative to the direct approach is to develop methods of approximating the absorption by bands or portions of bands by the use of band models. A large amount of effort and ingenuity has gone into the development of band models, and several are available. The choice among available models is dictated simply by the degree to which the model approximates experimental data, but, unfortunately, the data necessary to evaluate the models are not always available. Furthermore, a given model fits some bands or sections of bands better than others, and the goodness of fit depends on pressure, temperature, and optical thickness of the absorbing medium.

The simplest of the band models is that developed by Elsasser (1938) and consists of an infinite array of equally spaced Lorentz lines, all of equal intensity. An example of the Elsasser band is shown in Fig. 10.4A. The parameters which can be adjusted to optimize the fit to observational data are spacing, strength, and half-widths of the lines. It simulates the fundamental of the 15-μm carbon dioxide band reasonably well, as the lines in that case are somewhat regularly spaced, but there is little semblance of regularity in much of the water vapor and ozone spectra. In those cases, the Elsasser band would be expected to provide only a relatively gross approximation to reality.

The lack of regularity evident in much of the water vapor spectrum prompted development of a useful model which is variously termed the statistical, random, or Goody model. It was suggested independently by

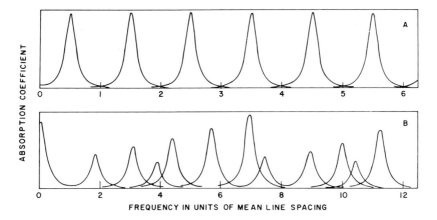

Fig. 10.4   Schematic diagrams of two absorption band models. (A) Elsasser model;
(B) statistical model with an exponential line intensity distribution. Note that the line
spacing is twice as great in (A) as in (B).

Mayer (1947) and Goody (1952). The main features of the statistical
model, a schematic diagram of which is shown in Fig. 10.4B, are a random
position and spacing of lines in the spectrum with some statistical distribution of line strength. The line strengths shown in Fig. 10.4B are for an
exponential distribution, but both the statistical form of the line-strength
distribution and the line spacing may be altered to fit observed conditions.
The line shape most frequently used in the statistical model is that of Lorentz, although the choice is somewhat arbitrary. The Doppler line shape
or a combination of Doppler and Lorentz shapes are probably more realistic
for the upper atmosphere.

   A number of other band models have been developed, but most are
modifications of these two. For instance, the random Elsasser model consists of a random superposition of different Elsasser bands with various line
strengths and spacings, and variations of the statistical model have included a constant line strength and line strengths following Gaussian or
other probability distributions.

   Useful as band models are, however, they have some basic limitations
associated with them. The main limitations are the following:

   1. Being statistical in nature, they apply to either entire bands or
considerable portions of bands and thus are capable of yielding only low
spectral resolution.

   2. There is always a statistical uncertainty associated with the results
obtained by the use of band models.

   3. Band models in general do not provide a method for accounting for

the effects of wings of bands located outside the interval under consideration.

4. Because of the variations of line half-widths with pressure, temperature, and absorber concentration, the optical thickness over slanted or vertical paths in the atmosphere is difficult to evaluate. The usual method of handling this nonhomogeneous case is to determine average value approximations of the optical thickness over the pathlength in question (see Section 10.5), but of course this further approximation introduces further complexity and uncertainties in the eventual results.

A comparison of absorption obtained by the use of two variations of the statistical band model with that for the Elsasser model, as given by Plass (1958), is shown in Fig. 10.5. In order to make such a comparison, it is necessary to adjust the line parameters of the distributions to give comparable total absorption. This was done for Fig. 10.5 by multiplying the average line strength for the statistical models by the factor $\pi/4$. The curves show that absorption for the statistical model with a constant line intensity is almost identical to that with an exponential line intensity distribution for all three values of $\beta = 2\pi\alpha/d$. Absorption for the Elsasser model is similar to that for the statistical model for the smaller optical thicknesses, but it is 10–15% greater in cases of high absorption.

The appropriate model for use on a given absorption band depends on the actual spacing and strengths of the individual lines, and the amount of

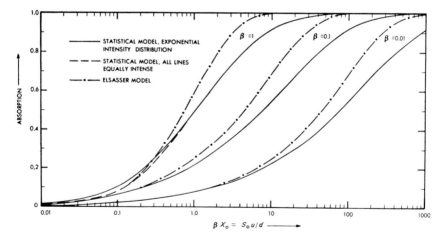

Fig. 10.5 Absorption as a function of the quantity S_0u/d for the Elsasser band model and two variations of the statistical model for three different values of $\beta = 2\pi\alpha/d$. (S_0, mean line intensity; u, amount of absorber; d, mean line spacing; α, half-width of the lines) (after Plass, 1958).

optical absorber in the path. Laboratory measurements show that the absorption for each model can be represented by certain approximations, depending on the amount of absorption near the line centers and the amount of overlapping of the wings of the lines. The following three useful approximations have been discussed by Plass (1958, 1960) among others.

Strong Line Approximation   If the absorption is virtually complete in the centers of the strongest lines in the band, then the band absorption can be represented as a function of only the linewidth α, line strength S, mean distance between the lines d, and mass of absorber u given by the quantity $\beta^2 x$, where $\beta = 2\pi\alpha/d$ and $x = Su/2\pi\alpha$. The strong line approximation is best for strong, wide, closely spaced lines and a large amount of absorber. The fractional absorption A is given for the Elsasser band model by the equation

$$A = \phi[(\tfrac{1}{2}\beta^2 x)^{1/2}] \tag{10.19}$$

where ϕ is the probability integral written as

$$\phi(z) = \frac{2}{\sqrt{\pi}} \int_0^z e^{-z^2}\, dz \tag{10.20}$$

and for the Goody (or statistical) model by

$$A = 1 - e^{-(2\beta^2 x/\pi)^{1/2}} \tag{10.21}$$

It is of interest that over small intervals of $\beta^2 x$, the absorption varies approximately as the quantity $(pu)^n$, where p is pressure, u is mass of absorber, and $0 \le n \le \tfrac{1}{2}$. For no overlapping of the lines, the square-root approximation ($n = \tfrac{1}{2}$) applies, whereas $n = 0$ implies strong overlapping of the lines.

Weak Line Approximation If the absorption is small everywhere, even in the centers of the strongest lines, the absorption is a function of only $\beta x = Su/d$, and is the same for all band models. This tends to explain the similarity of the curves for small absorptions shown in Fig. 10.5. In this case the fractional absorption is

$$A = 1 - e^{-\beta x} \tag{10.22}$$

Nonoverlapping Line Approximation   If the spectral lines do not overlap appreciably, the fractional absorption is well approximated in terms of the Bessel functions $J_0(x)$ and $J_1(x)$ by the expression

$$A = \beta x e^{-x}[J_0(x) + J_1(x)] \tag{10.23}$$

This relation is particularly useful at small values of u and p.

TABLE 10.1

*Criteria under Which the Various Approximations Yield Values of Fractional
Absorption Which Are in Error by Less Than 10%*

Approximation	$\beta = 2\pi\alpha/d$	Elsasser model	Statistical model 1	Statistical model 2
Strong line [Eqs.	10^{-3}	$x > 1.63$	$x > 1.63$	$x_0 > 2.4$
(10.20) and (10.21)]	10^{-2}	$x > 1.63$	$x > 1.63$	$x_0 > 2.4$
	10^{-1}	$x > 1.63$	$x > 1.63$	$x_0 > 2.3$
	1	$x > 1.35$	$x > 1.1$	$x_0 > 1.4$
	10	$x > 0.24$	$x > 0.24$	$x_0 > 0.27$
	100	$x > 0.024$	$x > 0.024$	$x_0 > 0.24$
Weak line [Eq. (10.22)]	10^{-3}	$x < 0.20$	$x < 0.20$	$x_0 < 0.10$
	10^{-2}	$x < 0.20$	$x < 0.20$	$x_0 < 0.10$
	10^{-1}	$x < 0.20$	$x < 0.20$	$x_0 < 0.10$
	1	$x < \infty$	$x < 0.23$	$x_0 < 0.11$
	10	$x < \infty$	$x < \infty$	$x_0 < \infty$
	100	$x < \infty$	$x < \infty$	$x_0 < \infty$
Nonoverlapping line	10^{-3}	$x < 600,000$	$x < 63,000$	$x_0 < 80,000$
[Eq. (10.23)]	10^{-2}	$x < 60,00$	$x < 630$	$x_0 < 800$
	10^{-1}	$x < 60$	$x < 6.3$	$x_0 < 8$
	1	$x < 0.7$	$x < 0.22$	$x_0 < 0.23$
	10	$x < 0.02$	$x < 0.020$	$x_0 < 0.020$
	100	$x < 0.002$	$x < 0.0020$	$x_0 < 0.0020$

Plass (1960) has tabulated the conditions under which the various
approximations yield values for the absorption which deviate from the
actual by less than 10%, as shown in Table 10.1. The statistical model 1
has all absorption lines of equal intensity, whereas for statistical model 2
the line intensity distribution has the exponential form

$$P(S) = \frac{1}{S_0} e^{-(S/S_0)} \tag{10.24}$$

Equation (10.21) is valid for this latter case if

$$x = \tfrac{1}{4}\pi x_0 = S_0 u/8\alpha \tag{10.25}$$

10.5 THE CURTIS–GODSON APPROXIMATION

Since pressure and temperature vary strongly in the atmosphere, the
half-widths and line intensities also vary with atmospheric conditions.

Such variations must be accounted for in determining the optical thickness. For a given line, the optical thickness τ_ν for optical mass u is given by the relation

$$\tau_\nu = \int K_\nu(p,\, T)\; du \tag{10.26}$$

where K_ν is a function of pressure and temperature.

The method most often applied to this nonhomogeneous case is that developed by Curtis (1952) and Godson (1953). By the Curtis–Godson approximation, the actual amount of absorbing matter is modified for temperature and pressure effects to give an effective amount of absorber at an effective pressure, after which the relations for the homogeneous case are applicable. The effective pressure p_e is given (Goody, 1964) by

$$p_e = \int p\bar{S}\; du \Big/ \int \bar{S}\; du \tag{10.27}$$

where p is pressure and $\bar{S}$ is mean line intensity. For the Elsasser and the random statistical band models, the effective optical mass is

$$u_e = \frac{1}{S_e} \int \bar{S}\; du \tag{10.28}$$

where S_e is an effective mean line intensity.

Equations (10.27) and (10.28) constitute the Curtis–Godson approximation for radiative transfer through inhomogeneous paths in the atmosphere. They give a useful simulation of reality in terms of an effective amount of absorber u_e at an effective pressure p_e. It has been found to be an extraordinarily good approximation for the strong line and weak line cases, and gives reasonably good results in the intermediate cases (Kaplan, 1959). Walshaw and Rodgers (1963) have found the approximation very good for the 15-μm band of CO_2 and reasonably good for the rotational band of water vapor. It does not give good results for the 9.6-μm band of ozone, a result which appears to be due to the increase of ozone with decreasing pressure in contrast to the usual case of decreasing optical mass with decreasing pressure. The approximation has been analyzed in considerable mathematical detail by Armstrong (1968). Increased accuracy may be obtained in case of a thick atmospheric layer by dividing it into several sublayers and applying the Curtis–Godson technique to the individual sublayers, as was done by Gates and Calfee (1966). Yamamoto et al. (1972) introduced a parameter for making the equivalent half-width smaller for intermediate pathlengths, for which case the original Curtis–Godson approximation seems to give too much absorption.

10.6   ABSORPTION SPECTRA OF ATMOSPHERIC GASES

Although there are many different gases in the atmosphere which absorb and emit radiation in the infrared spectral region, the three which are most important in determining the atmospheric energy budget are water vapor, carbon dioxide, and ozone. A brief description of their radiative characteristics is given below. More details are given in standard references (e.g., Goody, 1964), and an extensive bibliography on this and other atmospheric radiation problems has recently been issued by Howard and Garing (1971).

10.6.1   Water Vapor

Water vapor is the most abundant of the radiatively important gases in the Earth's atmosphere, and it has important absorption bands in both the solar and terrestrial wavelength ranges. The tropospheric temperature structure is maintained largely by the effects of water, both in its vapor form and as a liquid or solid in the cloud systems of the world. Available evidence indicates that there is comparatively little water vapor in the stratosphere, in which case its radiative effects would perhaps be of less significance than those of carbon dioxide and ozone, but chemical reactions between water vapor and ozone may cause significant decreases of the ozone concentration. This latter effect is of concern in assessing the possible environmental impact of aircraft flying at high altitudes in the atmosphere.

The water vapor molecule is a triatomic nonlinear molecule of the asymmetric-top type. It has a rich and complex vibration-rotation spectrum in the infrared. The ν_1 and ν_3 fundamental vibrations, involving stretching of the chemical bonds, are at too short wavelengths (2.74 and 2.66 μm, respectively) to be of much importance in the terrestrial regime, although they absorb significant amounts of solar radiation. The ν_2 fundamental due to bending of the bands is centered at 6.25 μm, and in combination with rotational transitions is responsible for the very strong 6.3 μm band, the wings of which extend from at least 5 to 9 μm. A pure rotational band of water vapor extends with varying intensity from about 18 to beyond 100 μm, with wings overlapping parts of the 15-μm band of CO_2. Both of these latter bands exert a strong influence on the radiative energy balance of the atmosphere.

A list of the principal bands of water vapor in the infrared region, as compiled by Goody (1964) from various sources, is given in Table 10.2.

10.6.2   Carbon Dioxide

Since the carbon dioxide molecule is of the linear type, CO_2 has a relatively simple absorption spectrum. Because of symmetry of the mole-

TABLE 10.2

Spectral Position of Centers, Approximate Limits, and Intensities of the More
Important Infrared Absorption Bands of Water Vapor[a]

Band name	Position of band center		Approximate band limits (μm)	Band intensity $cm^{-1} (atm\ cm)^{-1}$
	Wavelength (μm)	Wave number (cm^{-1})		
ρ	0.906	11032		0.05
	0.920	10869		0.01
σ	0.942	10613	0.870–0.990	0.3
	0.943	10600		0.02
τ	0.968	10329		0.05
	0.972	10284		0.001
—	1.016	9834		0.002
ϕ	1.111	9000		0.008
	1.135	8807		0.21
	1.141	8762	1.07–1.20	0.0003
	1.194	8374		0.0008
	1.209	8274		0.0002
ψ	1.343	7445		0.03
	1.379	7250		40.0
	1.389	7201	1.25–1.54	0.4
	1.455	6871		0.3
	1.476	6775		0.005
Ω	1.876	5331	1.69–2.08	5.9
	1.910	5235		0.2
—	2.143	4667	—	0.008
ν_3	2.662	3755.92	2.27–2.99	167.0
ν_1	2.734	3657.05		11.0
$2\nu_2$	3.173	3151.60	2.99–3.57	1.7
ν_2	6.270	1594.78	4.88–9.52	222.0

[a] Adapted from Goody, 1964.

TABLE 10.3

Band Intensities (at 300°K) for the Various Components of the 15-μm Band of CO_2 [a]

Band center (μm)	Band intensity (cm^{-2} atm^{-1})	Band center (μm)	Band intensity (cm^{-2} atm^{-1})
12.07	0.00022	14.97	16.5
12.64	0.0242	14.98	214.0
13.20	0.0099	15.45	0.77
13.48	0.158	13.87	5.5
13.54	0.0154	16.18	4.7
14.52	0.33	16.74	0.154
14.97	0.93	17.19	0.0092
		18.37	0.011

[a] Adapted from Drayson *et al.*, 1972.

cule, the ν_1 vibration has no spectrum at infrared wavelengths and no pure rotation spectrum for the $^{12}C^{16}O_2$ molecule. Various isotopic forms ($^{12}C^{16}O^{17}O$, $^{12}C^{16}O^{18}O$, etc.) do show a rotational structure, but their concentrations in the atmosphere are so low that they do not have a major influence on the overall terrestrial regime. The ν_2 fundamental centered at 14.98 μm, together with a number of overtone and combination bands, is responsible for the very strong 15-μm band of CO_2. Because of its position near the maximum of the Planckian radiation curves for atmospheric temperatures, the 15-μm band is very important in the terrestrial radiation regime. The intensities of the various bands which in combination make the 15-μm band, as tabulated by Drayson *et al.* (1972), are shown in Table 10.3.

The ν_3 bands centered near 4.3 μm are also very strong, but since they are in the region of low intensity of both terrestrial and solar radiation, they have a relatively minor influence in the energy exchange in the atmosphere. Other bands of CO_2 in the infrared region occur at about 5 μm and at about 9.4 and 10.4 μm, they are weaker than those mentioned above, and are of less importance in atmospheric radiation. The principal bands of CO_2 at the shorter infrared wavelengths as summarized from various sources by Drayson *et al.* (1972) are given in Table 10.4.

10.6.3   Ozone

The ozone molecule is of the triatomic nonlinear type with a relatively strong rotation spectrum. The three fundamental vibrational bands ν_1, ν_2, and ν_3 occur at wavelengths of 9.066, 14.27, and 9.597 μm, respec-

TABLE 10.4

*The Shorter Wavelength Bands of CO_2 Which Are of Importance
in the Terrestrial Radiation Regime*[a]

Region (μm)	Band center (μm)	Band intensity (cm^{-2} atm^{-1})
4.3	4.256	2700.0
	4.280	210.0
4.8	4.695	0.01
	4.777	0.41
	4.815	0.66
5.2	5.175	0.05
9.4	9.425	0.0159
10.4	10.41	0.0199

[a] Adapted from Drayson *et al.*, 1972.

tively. The location of the band centers and the band strengths for the
principal infrared bands of ozone, as tabulated by Drayson *et al.* (1972),
are given in Table 10.5.

The very strong ν_3 and moderately strong ν_1 fundamentals combine to
make the well-known 9.6-μm band of ozone. Since this band occurs in a
relatively good "window" for other gases and is located near the maximum
of the Planckian curve for atmospheric temperatures, it exerts a very significant influence on the infrared energy budget of the atmosphere, particularly at the higher altitudes. The ν_2 fundamental is well masked by the
15-μm band of CO_2, and the strong band at about 4.7 μm is in a weak portion of the Planckian energy distribution for the atmosphere.

TABLE 10.5

Location and Strengths of the Principal Infrared Bands of Ozone

Band center (μm)	Band strength (cm^{-2} atm^{-1})	Band center (μm)	Band strength (cm^{-2} atm^{-1})
9.597 (ν_3)	350.0	5.580	0.54
9.066 (ν_1)	10.4	3.598	0.66
4.739	32.0	3.287	3.0
14.27 (ν_2)	18.0	3.144	0.33
5.787	1.35		

10.6.4 Trace Gases

The radiatively active trace gases in the atmosphere are HDO (heavy water), nitrous oxide, methane, carbon monoxide, ammonia, and a few others. Although some of them have very significant absorptions in the shorter portions of the solar spectrum, their effects in the terrestrial spectrum are minor compared to those of water vapor, carbon dioxide, and ozone. The most important of the bands of the trace gases at wavelengths greater than 4 μm are listed in Table 10.6.

Heavy water (HDO) constitutes less than 0.03% of the water vapor in the atmosphere. The HDO molecule has quite different vibrational frequencies from those of ordinary water, and the bands are well separated. The small relative quantity of HDO, however, makes its radiative effects of minor significance in overall energy considerations.

The absorption spectrum of nitrous oxide is very rich in the vacuum ultraviolet. In the infrared region there are several moderately strong bands arising from the three fundamental vibrations, plus a rotational spectrum due to the asymmetric configuration of the N_2O molecule, and a number of overtone, combination, and upper-state bands.

The only methane band of note in the terrestrial spectrum is the ν_4 fundamental centered at 7.66 μm. There are, however, a number of overtone and combination bands in the solar infrared region between 1.67 and 3.85 μm.

Carbon monoxide exhibits some electronic transitions in the ultraviolet and some rotational lines in the millimeter wavelength range, neither

TABLE 10.6

Principal Absorption Bands of the Trace Gases for Wavelengths Greater Than 4 μm

Gas	Wavelength (μm)
HDO	7.126
N_2O	4.06
	4.50 (ν_3)
	4.52
	7.78 (ν_1)
	9.56
	17.0 (ν_2)
CH_4	7.66 (ν_4)
CO	4.67

of which is of much significance in atmospheric studies. The only band in the 4–100 μm region is the single fundamental at 4.67 μm.

10.7   COMPUTATIONS OF FLUXES AND HEATING RATES

The determination of fluxes and rates of heating or cooling for real atmospheric cases involves the following five specific steps, although the order of the steps may be changed to accommodate specific computational methods.

1. Development of a formal solution to the equation of radiative transfer (see Section 10.2). For this purpose, a plane parallel atmosphere and isotropic source function are usually assumed, although more general cases can be handled.
2. Integration over zenith angle to obtain monochromatic radiative fluxes instead of intensities.
3. Integration over some height coordinate (altitude, pressure, or absorber amount) to obtain monochromatic fluxes at discrete heights. This step implicitly involves approximations (see the Curtis–Godson approximation, Section 10.5) for changing line properties with changing pressure, temperature, and absorber amount.
4. Integration over spectral interval to get total energy flux.
5. Differentiation of flux distribution with respect to height to obtain flux divergence for computing rates of heating or cooling of the atmosphere.

A method must be provided in any computational scheme to introduce the appropriate boundary conditions, the boundaries of most interest being the underlying ground or water surface, the surfaces of clouds at various altitudes, and the effective top of the atmosphere.

A number of different techniques for determining fluxes and heating rates have been developed, some of the more classical of which are listed in Table 10.7. The Elsasser chart has enjoyed extensive use in the United States, and blank charts may be purchased from the American Meteorological Society. Relatively complete and useful discussions of the Elsasser chart, Yamamoto chart, and Shekhter chart have been given by Sellers (1965), Goody (1964), and Kondratyev (1969), respectively.

The advent of large and fast electronic computers ushered in the possibility of a new generation of techniques for computing heating and cooling of the atmosphere, and more recently, the development of numerical methods of weather forecasting opened up new requirements for radiative input data. As a result of these advances, many authors, including Curtis

TABLE 10.7

Methods and Techniques for Determining Flux and Flux Divergence in the Atmosphere

Method	Type	Quantity
Moller (1944)	Graphical	Flux
Elsasser (1942)	Graphical	Flux
Dimitriyev and Narovlansky (1944)	Graphical	Flux
Robinson (1947, 1950)	Graphical	Flux
Shekhter (1950)	Graphical	Flux
F. A. Brooks (1952)	Graphical	Flux
Yamamoto (1952)	Graphical	Flux
Niilisk (1961)	Graphical	Flux
D. L. Brooks (1950)	Numerical	Flux
Elsasser and Culbertson (1960)	Numerical	Flux
Bruinenberg (1946)	Numerical	Flux divergence
Hales (1951)	Graphical	Flux divergence
Yamamoto and Onishi (1953)	Graphical	Flux divergence

(1956), Murgatroyd and Goody (1958), Kaplan (1959), Davis and Viezee (1964), Kuhn (1966), Rodgers and Walshaw (1966), Drayson (1967), Yamamoto *et al.* (1972), Arking and Grossman (1972), and others have developed improved methods for computing radiative transfer in the infrared region. In addition, the very important problem of the determination of temperature and gas absorber profiles of the atmosphere from satellite radiation measurements has produced a great deal of work in atmospheric transmissivity theory and measurements (e.g., Wark and Fleming, 1966; Smith, 1970; Conrath, 1972; Chahine, 1974). Unfortunately the radiometric inversion problem is outside the scope of this book.

One of the most complete of these computer based methods is that of Rodgers and Walshaw (1966). It is applicable for computations at all levels in the atmosphere, it includes the effect of temperature on line intensity, and the influences of the various sources of error are evaluated. Reference is made to the original paper for the details, but the general outline of the method is given below.

Beginning with the basic radiative transfer equation [see Eq. (10.4)] applicable to any height z in a plane parallel model of the atmosphere, the steps in the Rodgers—Walshaw process are (1) cast the equations for upward and downward fluxes in terms of a flux transmission function, (2) differentiate the equations with respect to height and combine them to obtain a formal expression for the atmospheric cooling rate, (3) integrate the equations with respect to height to obtain the cooling rate at height z for a spectral interval sufficiently narrow for the Planck function to be sensibly constant, (4) integrate the results with respect to frequency to obtain a mean transmissivity for the entire spectrum, (5) use the Curtis–Godson approximation to take account of variable pressure, temperature, and absorber amount as a function of height, and (6) integrate over zenith angle to get the quantities in terms of flux and flux divergence instead of monodirectional intensities. These steps will be discussed in order.

Fluxes at Level z If we define a transmission function $T_r(z, z')$ applicable to a narrow spectral interval r for flux (instead of monodirectional intensity) through the atmosphere between levels z and z', then we can write expressions for the upward and downward fluxes in the form

$$F_r^+(z) = T_r(z, 0) B_r(g) + B_r(z) - \int_0^z T_r(z, z') \frac{dB_r(z')}{dz'} dz'$$

$$(10.29)$$

$$F_r^-(z) = T_r(z, Z) B_r(Z) - B_r(z) - \int_z^Z T_r(z, z') \frac{dB_r(z')}{dz'} dz'$$

Here B_r is the Planck function in units of flux at the temperature of the ground g or atmosphere at level z, z', or Z, the last being the highest level encompassed by the computation.

Atmospheric Cooling Rate The cooling rate (time rate of loss of energy per unit volume) for spectral interval r is obtained by differentiating Eqs. (10.29) with respect to height and combining them to give

$$C_r(z) = \frac{dT_r}{dz}(z, Z) B_r(Z) - \frac{dT_r(z, g)}{dz} B_r(g)$$

$$- \int_0^Z \frac{dT_r(z, z')}{dz} \frac{dB_r(z')}{dz'} dz' \qquad (10.30)$$

The three terms on the right-hand side represent the divergence at level z of fluxes from the upper boundary, lower boundary, and from all levels z' between the boundaries.

Integration with Respect to Height A direct integration of Eq. (10.30) with respect to height is not feasible in general because of difficulties in determining $dT_\nu(z, z')/dz$. A more practical method for use in actual cooling-rate calculations is that developed by Curtis (1956), and outlined as follows. A set of n reference levels at which the cooling rate is to be calculated are defined, and the temperature and absorber amount are determined at these levels by interpolation between the observational datum points. The one constraint on the choice of levels is that they be equally spaced in a relevant parameter. Curtis used equal intervals of $\log_{10}\alpha$ where α is Lorentz line-width, while Rodgers and Walshaw used equal intervals of $\ln(p/p_0)$, p_0 and p being pressure at the ground and at the reference level, respectively. By approximating the distribution of the Planck function between the layers by some type of interpolation formula, then the rate of cooling at level z_k due to radiation from the other j reference levels in spectral interval r can be written as

$$C_r(z_k) = \sum_{j=1}^{n} A_{kj}B_r(z_j) \qquad (10.31)$$

Matrix elements A_{kj} are dependent on the distribution and type of absorber, and are thus closely related to the transmissivity. Rodgers and Walshaw point out that if the temperature dependence of absorption is small, then for a given distribution of absorber the elements A_{kj} can be calculated independently of $B_r(z_j)$ and used for any temperature profile without introducing errors greater than about 0.2°C per day. This procedure would be advantageous for carbon dioxide, but water vapor and ozone have such variable concentrations that A_{kj} will have to be computed for each case. Thus the precalculation of A_{kj} would be of restricted value for general use, and the more complete calculation should present no problem for modern computers.

The method of interpolation of B_r between levels is mainly a matter of convenience. Rodgers and Walshaw used a cubic interpolation formula, while Drayson (1967) used a fourth-degree polynomial for layers near to level z_k and a quadratic for more distant layers.

Integration over Frequency By use of the statistical band model and Lorentz line shape, the mean transmissivity over a spectral region for which the model is applicable is given by

$$\bar{T} = \exp[-(Su/d)(1 + Su/\pi\alpha)^{1/2}] \qquad (10.32)$$

S, d, and α being mean intensity, spacing, and half-width of the Lorentz lines and u being absorber amount. Other band models could be used, but Rodgers and Walshaw found that this model gives accuracies commensurate with uncertainties in the observational data.

The Curtis-Godson Approximation This approximation, by which effects of changing pressure along a pathlength are taken into account, was discussed above [Eq. (10.27)]. An extension of the method, contained in an unpublished communication from Godson, was used by Rodgers and

Walshaw (1966) for introducing corrections to line intensity S and half-width α for changing temperatures along the atmospheric pathlength. We define the temperature-dependent functions $\Phi(T)$ and $\Psi(T)$ in terms of ratios of the quantities at any temperature T and a standard temperature T_0 as

$$\Phi(T) = \sum_{i=1}^{n} [S_i(T)/S_i(T_0)]$$

$$\Psi(T) = \left\{ \sum_{i=1}^{n} [S_i(T)\alpha_0(T)]^{1/2} \Big/ \sum_{i=1}^{n} [S_i(T_0)\alpha_0(T_0)]^{1/2} \right\}$$

(10.33)

where the sum is taken over the n lines in the frequency interval. Then the modified absorber mass $\bar{u}$ and average pressure $\bar{p}$ are given by the relations

$$\bar{u} = \int \Phi(T)\, du$$

$$\bar{u}\bar{p} = \int \Psi(T)p\, du$$

(10.34)

The improvement of accuracy achieved by this additional correction is still largely untested in actual atmospheric cases. In one case of cooling calculations for the water vapor bands, Rodgers and Walshaw (1966) found that the approximation using temperature corrections reproduced the results obtained by the full calculation to better than 1%. A rigorous evaluation of Eqs. (10.33), involving a summation over all of the relevant lines, would be very demanding in computer time, and not all of the required data on spectral lines are available. Perhaps a simple empirically derived correction can be used to replace the detailed computation and still retain the main improvements of the method. One such relation is given by Rodgers and Walshaw, as a series expansion in terms of temperature differences from a standard.

Integration over Zenith Angle   The final step necessary to determine fluxes and rates of heating or cooling is to integrate the transmissivity function over zenith angle θ to obtain the flux transmissivity of Eqs. (10.29) and (10.30). In other words, we need for each frequency interval r the quantity

$$T_r(z, z') = 2 \int_0^{\pi/2} \left[\frac{1}{\Delta\nu} \int_{\Delta\nu} T_\nu(z, z', \theta)\, d\nu \right] \sin\theta \cos\theta\, d\theta \quad (10.35)$$

The usual method of evaluating such an integral is by use of the very convenient properties of the exponential integral (see Elsasser and Culbertson, 1960; Drayson, 1967), or by some method of numerical quadrature in conjunction with an electronic computer. In the Rodgers–Walshaw method the angular integration is performed after the frequency integration and the Curtis–Godson approximation have been introduced, in which case the angular integration is not as simple as indicated by Eq. (10.35). Since the Rodgers–Walshaw development is lengthy, we give only the final result, which is itself a somewhat complicated expression but not difficult to evaluate. The quantity dT/dz of Eq. (10.30) for the statistical band model is

$$\frac{dT_r}{du}(z, z') = \frac{1}{\bar{u}\bar{p}}\left[\Phi\bar{p}M_1 + \Psi\bar{p}M_2\right] \tag{10.36}$$

For the statistical model with a Lorentz line shape and the Curtis–Godson temperature-dependent approximation, the parameters M_1 and M_2 are given as

$$M_1 = \pm 2 \int_0^1 \frac{abx^{3/2}}{c^{3/2}} \exp\left[-\frac{ab}{(cx)^{1/2}}\right] dx$$
$$M_2 = \pm 2 \int_0^1 \frac{a^2bx^{1/2}}{c^{3/2}} \exp\left[-\frac{ab}{(cx)^{1/2}}\right] dx \tag{10.37}$$

where $a = Su/2\pi\alpha_0$, $b = 2\pi\alpha_0/d$, and $c = x + 2a$.

In the interests of convenience, it should be mentioned that a very simple approximation for the angular integration was suggested by Elsasser (1942). This relationship gives the flux transmissivity with the basic Curtis–Godson approximation as

$$T(z, z') = T_b(r, \bar{u}, \bar{p}) \tag{10.38}$$

where T_b represents monodirectional beam transmissivity and r is a constant which is usually termed the diffusion factor. The exact value of r is a matter of some uncertainty (see Goody, 1964). Elsasser (1942) gave the value 1.66, while Elsasser and Culbertson (1960) reduced it to 1.60. In testing the validity of the use of the diffusion factor, Rodgers and Walshaw (1966) obtained the amazingly small relative maximum error in the cooling rate of 1.5%, their computations being on the basis of $r = 1.66$.

The validity of any method of computing heating or cooling rates must be finally evaluated by comparison with observations. Unfortunately, the required observational data are difficult to obtain and are not yet generally available. Thus the best that can be done at this time is to compare results obtained by the different methods from the same atmospheric data. Such comparisons have been made by Kondratyev and Niilisk (1966) and by Rodgers and Walshaw (1966). The data of Kondratyev and Niilisk show differences as large and 30–50% in both fluxes and heating rates among the various methods. These large variations must be viewed as an indication of the accuracy obtainable with radiation charts or numerical computations. A comparison of results obtained with an Elsasser chart and those from measurements of the flux of radiation downward on a horizontal surface given by Morgan *et al.* (1970) from a series of eight different determinations under clear skies at Davis, California indicates that somewhat better accuracy than this may be attainable under selected conditions. Morgan *et al.* show a standard êrror of estimate of 0.009 for a mean flux of 0.403, or only about 2%. The correlation coefficient between the two sets of data was 0.985.

The comparison by Rodgers and Walshaw for five different methods of computation is shown in Fig. 10.6. If the Rodgers–Walshaw results are taken as completely accurate, the methods of Brooks and of Yamamoto and Onishi give too much cooling everywhere, and that of Elsasser too little cooling in the troposphere and too much in the stratosphere. Of course, no claim has been made by Rodgers and Walshaw for high accuracy in their results, so conclusions regarding comparative accuracies are not well founded. To the extent that the comparisons of Morgan *et al.* (1970) are

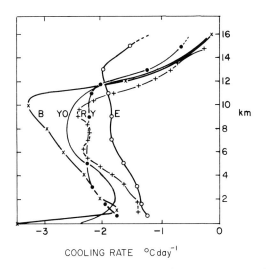

Fig. 10.6 Comparison of cooling rates versus altitude as obtained by various authors from the same atmospheric data (after Rodgers and Walshaw, 1966). E, Elsasser (1942); B, Brooks (1950); Y, Yamamoto (1952); YO, Yamamoto and Onishi (1953); R, Rodgers and Walshaw (1966).

valid, the close agreement between the Elsasser and Rodgers–Walshaw methods in the low levels would tend to lend support to the latter.

REFERENCES

Arking, A., and Grossman, K. (1972). The influence of line shape and band structure on temperatures in planetary atmospheres. *J. Atmos. Sci.* **29**, 937–949.

Armstrong, B. H. (1968). Analysis of the Curtis-Godson approximation and radiation transmission through inhomogeneous atmospheres. *J. Atmos. Sci.* **25**, 312–322.

Brooks, D. L. (1950). A tabular method for the computation of temperature change by infrared radiation in the free atmosphere. *J. Meteorol.* **7**, 313–321.

Brooks, F. A. (1952). Atmospheric radiation and its reflection from the ground. *J. Meteorol.* **9**, 41–52.

Bruinenberg, A. (1946). Een numericke methode voor de bepaling van temperatuurs—veranderingen door straling in de vrije atmosfeer. *K. Ned. Meteor. Inst. Meded. Verh. Ser. B.* **1** (1), 52.

Chahine, M. T. (1974). Remote sounding of cloudy atmospheres: I. The single cloud layer. *J. Atmos. Sci.* **31**, 233–243.

Conrath, B. J. (1972). Vertical resolution of temperature profiles obtained from remote radiation measurements. *J. Atmos. Sci.* **29**, 1262–1271.

Curtis, A. R. (1952). Discussion. *Quart. J. Roy. Meteorol. Soc.* **78**, 638–640.

Curtis, A. R. (1956). The computation of radiative heating rates in the atmosphere. *Proc. Roy. Soc. Sect. A*, **236**, 148–156.

Davis, P. A., and Viezee, W. (1964). A model for computing infrared transmission through atmospheric water vapor and carbon dioxide. *J. Geophys. Res.* **69**, 3785–3794.

Dimitriyev, A. A., and Narovlansky, G. I. (1944). The calculation of long–wave radiant fluxes coming from a limited solid angle and from a semiinfinite atmosphere. *Proc. Acad. Sci. USSR, Geograph. Geophys. Ser.*, **6**.

Drayson, S. R. (1967). The calculation of long–wave radiative transfer in planetary atmospheres. Tech. Rept., Grant GP-4385, Dept. Meteorology and Oceanography, Univ. of Michigan, Ann Arbor, Michigan.

Drayson, S. R., Bartman, F. L., Kuhn, W. R., Tallamraju, R. (1972). Satellite measurement of stratospheric pollutants and minor constituents by solar occultation: A preliminary report. Final Rept., Grant NG-10-72, High Alt. Eng. Lab., Univ. of Michigan. Ann Arbor, Michigan.

Elsasser, W. M. (1938). Mean absorption and equivalent absorption coefficient of a band spectrum. *Phys. Rev.* **54**, 126–129.

Elsasser, W. M. (1942). "Heat Transfer by Infrared Radiation in the Atmosphere," Harvard Meteorol. Studies. No. 6. Harvard Univ. Press, Cambridge, Massachusetts.

Elsasser, W. M., and Culbertson, M. F. (1960). "Atmospheric Radiation Tables." Meteorol. Monographs, Vol. 4, No. 23. Amer. Meterol. Soc., Boston, Massachusetts.

Gates, D. M., and Calfee, R. F. (1966). Calculated slant-path absorption and distribution of water vapor, *Appl., Opt.*, **5**, 287–292.

Godson, W. L. (1953). The evaluation of infrared radiative fluxes due to atmospheric water vapour, *Quart. J. Roy. Meteorol. Soc.* **79**, 367–379.

Goody, R. M. (1952). A statistical model for water-vapour absorption, *Quart. J. Roy. Meteorol. Soc.* **78**, 165–169.

Goody, R. M. (1964). "Atmospheric Radiation I: Theoretical Basis." Oxford Univ. Press (Clarendon), London and New York.

Hales, J. V. (1951). An atmospheric radiation flux divergence chart and meridional cross sections of water vapor radiational heat losses computed through its use. Ph.D. Dissertation, Dept. Meteorol., Univ. California, Los Angeles, California.

Hitchfield, W., and Houghton, J. T. (1961). Radiative transfer in the lower stratosphere due to the 9.6 micron band of ozone. *Quart. J. Roy. Meteorol. Soc.* **87**, 562–577.

Howard, J. N., and Garing, J. S. (1971). Atmospheric optics and radiative transfer. *Trans. Amer. Geophys. Union* **52**, 371–389.

Kaplan, L. D. (1959). A method for calculation of infrared flux for use in numerical models of atmospheric motion. *In* "The Atmosphere and the Sea in Motion," pp. 170–177. Rockefeller Institute and Oxford Univ. Press, New York and London.

Kondratyev, K. Ya. (1969). "Atmospheric Radiation." Academic Press, New York.

Kondratyev, K. Ya., and Niilisk, H. Y. (1960). Comparison of radiation charts. *Geofis. Pura Apl.* **46**, 231.

Kuhn, W. R. (1966). Infrared radiative transfer in the upper stratosphere and mesosphere. Sci. Rept., Dept. of Astro-Geophysics, Univ. of Colorado, Boulder, Colorado.

Mayer, H. (1947). Los Alamos Scientific Lab. Report No. LA-647.

Morgan, D. L., Pruitt, W. O., and Lourence, F. J. (1970). Radiation data and analysis for the 1966 and 1967 Meteorological Field Runs at Davis, California. Rept. TR ECOM 68-G10-2, Univ. of California, Davis, California.

Moller, F. (1944). Grundlagen eines diagrammes langwelliger strahlungsstrome. *Meteorol.*

Murgatroyd, R. J., and Goody, R. M. (1958). Sources and sinks of radiative energy from 30 to 90 km., *Quart. J. Roy. Meteorol. Soc.* **87**, 125–135.

Niilisk, H. Y. (1961). A new radiation chart. *Proc. Esthonian Acad. Sci., Ser. Phys. Math,* **10**, No. 4, 329–339.

Plass, G. N. (1958). Models for spectral band absorption. *J. Opt. Soc. Amer.* **48**, 690–703.

Plass, G. N. (1960). Useful representations for measurements of spectral band absorption. *J. Opt. Soc. Amer.* **50**, 868–875.

Robinson, G. D. (1947). Notes on the measurement and estimation of atmospheric radiation—1. *Quart. J. Roy. Meteorol. Soc.* **73**, 127–150.

Robinson, G. D. (1950). Note on the measurement and estimation of atmospheric radiation—2. *Quart. J. Roy. Meteorol. Soc.* **76**, 37–51.

Rodgers, C. D., and Walshaw, C. D. (1966). The computation of infrared cooling rate in planetary atmospheres. *Quart. J. Roy. Meteorol. Soc.* **92**, 67–92.

Sellers, W. D. (1965). "Physical Climatology." Univ. of Chicago Press. Chicago, Illinois.

Shekhter, F. N. (1950). Calculation of thermal radiant fluxes in the atmosphere. *Trans. Main Geophys. Obs., Leningrad,* **22**.

Smith, W. L. (1970). Iterative solution of the radiative transfer equation for the temperature and absorbing gas profile of an atmosphere. *Appl. Opt.* **9**, 1993–1999.

Walshaw, C. D., and Rodgers, C. D. (1963). The effect of the Curtis-Godson approximation on the accuracy of radiative heating-rate calculations. *Quart. J. Roy. Meteorol. Soc.* **89**, 122–130.

Wark, D. Q., and Fleming, H. E. (1966). Indirect measurements of atmospheric temperature profiles from satellites. *Mon. Weather. Rev.* **94**, 351–362.

Yamamoto, G. (1952). On a radiation chart. Sci. Rept. Ser. 5 (Geophys.), **4**, 9–23, Tohoku Univ., Japan.

Yamamoto, G., and Onishi, G. (1953). A chart for the calculation of radiative temperature changes. Sci. Rept. Ser. 5 (Geophys.), **4**, 108–115. Tohoku Univ., Japan.

Yamamoto, G., Aida, M., and Yamamoto, S. (1972). Improved Curtis-Godson approximation in a non-homogeneous atmosphere. *J. Atm. Sci.* **29**, 1150–1155.

CHAPTER ELEVEN

Terrestrial Radiation: Methods of Measurement

11.1 MEASUREMENT OF LONG-WAVE RADIATION

The problems and techniques of measuring long-wave (or terrestrial) radiation are somewhat different from those for measurements of short-wave (or solar) energy. Since natural land and water surfaces, gases of the atmosphere, clouds, and similar materials on Earth are at much lower temperatures than that of the Sun, the wavelengths at which they emit significant amounts of energy are longer than the wavelengths of most solar energy. Almost all terrestrial emission is at wavelengths in the 4–100 μm region, whereas most solar energy is confined to wavelengths below 4 μm. In the terrestrial regime the entire environment, including the instrument itself, is emitting radiation at the same wavelengths and of comparable intensities as those of the radiation to be measured. The situation is in many respects comparable to the hypothetical case of trying to determine the intensity of a light source in a brightly lighted room with an instrument which is illuminated inside and out for which the detector itself is luminous.

Another difficulty in terrestrial radiation measurements is that the properties of real materials are not ideal. In the solar regime, one can use glass or quartz to isolate spectral ranges and to protect the sensing element, since these materials transmit well throughout the wavelength range of most of the solar energy and are inexpensive and easy to fabricate. As will be seen below, no comparable materials exist for the terrestrial wavelength range. The absorptivity of paints is in general wavelength dependent, and the sensors are sensitive to solar radiation as well as to terrestrial radiation.

As a result of difficulties of measurement, and perhaps a lack of interest in the early days, observations in the terrestrial radiation regime have been pursued mainly since World War II. Before that time, the instruments capable of measuring terrestrial radiation were few in number, the main ones being those of Ångström (1905), the Smithsonian Institution (Abbot *et al.*, 1922), and Albrecht (1933a, 1933b), and as far as is known no serious attempt had been made to set up a network of observations for energy-balance studies. Most of what was known about the terrestrial radiation balance and radiation energy profiles in the atmosphere had been deduced from theory. Since about 1946 considerable interest in the radiation energy balance has been generated, resulting in several new types of instruments being devised and the initiation of some systematic series of measurements.

Even with the new techniques, however, the interest has not been all one could wish. The number of stations at which long-wave radiation, either hemispherical or net, is measured scarcely compares with the number of solar radiation stations. During the 1957–1959 time period of the IGY, only 35 stations worldwide reported surface measurements of terrestrial radiation during the whole of 1958, with another 15 reporting partial records (Robinson, 1964). The majority of the stations were operated by the U.S.S.R., the United Kingdom, and New Zealand. Only one station in the continental U.S. reported net radiation balance for the 1958 period (Natick, Mass.). One U.S. reporting station was on an ice island in the Arctic, and the Antarctic included two stations operated entirely by the U.S. and one operated jointly with New Zealand. Most of the long-wave instruments during the IGY were of the net radiation type. The accuracies attained, as quoted by Robinson (1964), are the following:

Hourly sums:	$\pm (20\% + 5 \text{ cal cm}^{-2})$
Daily sums:	$\pm (15\% + 10 \text{ cal cm}^{-2})$
Monthly sums:	$\pm (15\% + 100 \text{ cal cm}^{-2})$
Annual sums:	$\pm (15\% + 200 \text{ cal cm}^{-2})$

The observation situation has not improved much in the U.S. since the time of the IGY. There is still no regularly reporting network for terrestrial radiation. Data from only one station (Palmer, Alaska) are reported in the National Summary of Climatological Data, and those data are of an experimental nature. In contrast, about 80 stations in the U.S. report solar radiation data. The most extensive network for terrestrial radiation measurements is that of the U.S.S.R., which has been expanded to about 100 stations reporting net radiation data on a daily basis (Budyko,

1972). The networks of some other countries have also been expanded since 1958, but the details are not immediately available.

As is the case with solar radiation instruments, considerable confusion has developed in the terminology for instruments measuring terrestrial radiation. In order to not add to the confusion, we shall use the definitions given by the World Meteorological Organization and listed above in Chapter 1. Briefly, an instrument for measuring the total radiation (solar plus terrestrial) in one direction through a horizontal surface is a pyrradiometer, while that for measuring the difference of total radiation in the upward and downward directions is a net pyrradiometer. In the literature the first of these has been variously termed effective pyranometer or hemispherical radiometer, while the second is sometimes designated a radiation balance meter or a net radiometer. A pyrgeometer is the same as a pyrradiometer except that it is used for measuring only terrestrial radiation (e.g., for measurements at night).

Several other types of radiometers sensitive to long-wave radiation have been developed for special purposes. Among the other types are the bolometer, directional radiometers, "black-ball" radiometers, radiation thermometers, satellite radiometers of various types, and infrared mapping devices. Because of space limitations these will not be discussed here. Instead, the emphasis will be on instruments for measuring total or net radiative fluxes on a horizontal surface.

Criteria by which a pyrradiometer or net pyrradiometer must be judged are not much different from those of meteorological instruments in general, but some are unique to the radiation problem and some are more difficult to meet in long-wave radiation instruments than in other types of devices. The most important criteria are the following.

Reliability This is undoubtedly the most important criterion for an instrument in continuous use. A reliable radiometer is one which gives reproducible results in sequential measurements, and for an operational device the reproducibility must be maintained over a long period of time.

Accuracy Reasonable accuracy in radiation instruments is very difficult to achieve and to maintain in an operational setting. To obtain the desired accuracy, the instrument must be properly calibrated under known conditions, instrument characteristics must be stable, and response to changes in environmental conditions must be either constant or known to within the limits of error required in the measurements.

Sensitivity The output signal must be large enough to be recorded by conventional means. Modern recorders and electronics techniques, however, are so advanced that sensitivity is not the problem it once was.

Simplicity and Convenience   The day-in day-out use of any instrument virtually demands that it be convenient to use, and radiometers are no exception. In addition, the possible lack of technical skill in an operational setting calls for simplicity of design and routine maintenance of the equipment.

Speed of Response   No goals have been established for the response time of continuously operated radiometers. For hourly or daily sums of radiation fluxes, a long coefficient of a minute or two is certainly adequate except in very extreme situations. Any errors due to instrument lag are probably small in comparison to those arising from other causes in climatological measurements. For airborne instruments, however, response time is of critical importance in obtaining a faithful reproduction of radiative profiles.

Durability   The weathering of radiometers is a serious problem, particularly for those in which the sensing element is directly exposed to dust, rain, and other elements of the environment. The sensing surfaces must be of a material which is subject to a minimum of deterioration.

Spectral Characteristics   No material has yet been found which has the same radiative characteristics at all wavelengths. This presents a problem for which no complete solution has been found in instruments required to measure fluxes of both solar and terrestrial radiation. Window materials absorb selectively; polished surfaces reflect selectively. As will be seen below, such selectivity is a source of considerable error, and efforts to take it into account have been only marginally successful.

Angular Response   The failure of an instrument to respond according to the cosine law as the angle of incidence of radiation on the sensor changes is a universal source of error in hemispheric type radiometers. This "cosine effect," which arises mainly from paints covering the surfaces of sensors, has not been well evaluated in most existing instruments, but it is a source of considerable error for which corrections are difficult to apply.

Other criteria for judging radiometers are portability, adaptability for remote and unattended operation, and cost, but these are normally evaluated according to the specific application for which the device is to be used.

11.2   EARLY TYPES OF LONG-WAVE RADIOMETERS

Boys Radiomicrometer   This device, developed by Sir Charles Boys (1885–1944), consists of an antimony–bismuth thermocouple connected to the coil of a mirror galvanometer suspended between two poles of a permanent magnet. Radiation falling on the thermojunction produces a current in the coil, resulting in a deflection of the galvanometer. The deflection is a function of the radiant energy absorbed by the thermocouple.

Radio Balance of Callendar   Two versions of the radio balance were developed by H. L. Callendar (1863–1930) beginning about 1900 (Callendar, 1910). Both are based on the principle of compensation of the heating caused by absorption of radiation by cooling produced the thermoelectric (Peltier) effect. The main difference between them is the shape of the radiation receiver, one being in the form of a disk and the other a cup. The disk type, the receiver of which was a blackened disk of 3 to 4 mm in diameter and 3 to 5 mm thick, was simple to construct, but the absorptivity of the paint was not well known, and the use of a single thermocouple limited its sensitivity. The cup form, however, had the advantage of essentially complete absorptivity, and it provided the possibility of incorporating several thermocouples for higher sensitivity. A schematic diagram of the cup-type device is shown in Fig. 11.1, in which A is a sensitive ammeter measuring the current in the Peltier circuit, B is a battery, and R is a control rheostat. G is a null galvanometer attached to thermocouples C and F for assuring equality of temperature of the two cups.

Ångström Pyrgeometer   The pyrgeometer of K. Ångström (1905) uses the electric compensation principle developed by him for the Ångström pyrheliometer (see Section 3.2.3). The sensor consists of two blackened

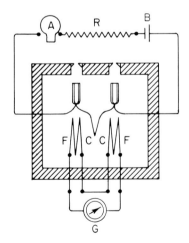

Fig. 11.1 Schematic diagram of the cup radiobalance (see text for explanation of labels). (After Callendar, 1910). Published by permission of The Institute of Physics and the Physical Society.

and two polished manganin resistance strips. Since the emissivity of
the polished strips is very low, they assume the same temperature as that
of the air. The blackened strips, having a high emissivity, exchange
radiation freely with the surroundings and generally cool down in nighttime conditions. The net loss of heat is compensated by electrically heating
the blackened strips. Equivalence of temperature of blackened and polished strips is determined by thermocouples attached to their reverse
sides. This gives a measure of the difference between the downward flux of
radiation absorbed and the upward flux of radiation emitted by the
blackened surface. The upward flux can be calculated from the Stefan–
Boltzmann law for the measured temperature of the strip, from which the
downward flux can be determined.

The Ångström pyrgeometer is suitable for use in the open air only at
night under relatively calm conditions. Appreciable wind circulation
causes nonradiative heat transfers which are variable and difficult to
evaluate. The instrument is sometimes used as a secondary reference for
measurements in the laboratory, after being calibrated against a primary
standard. Great care must be exercised to assure equilibrium temperature
conditions inside the instrument before radiation measurements are made
(CSAGI, 1958). The device is not suitable for automatic recording of
its output.

Albrecht Pyrradiometer and Net Pyrradiometer   The first of the instruments
(Albrecht, 1933a) consisted of a flat blackened metallic plate, to the back
of which was attached a 50-junction manganin-constantan thermopile.
The plate could be heated by an electric current through platinum resistance wires in thermal contact with the plate to provide a correction for
effects of the wind. The assembly was protected from radiation from
below by a second plate of polished metal. One measurement of incident
radiation required two observations at different heating rates H_1 and H_2.
If the corresponding plate temperatures are T_1 and T_2, then the flux of
radiation incident on the plate is given by the relation

$$F = \frac{T_1}{T_2 - T_1}\,(H_2 - H_1) - H_1 \qquad (11.1)$$

In practice, two of the devices were used with different heating rates. This
permitted recording of the outputs and thus obtaining a continuous record
of the incident flux.

The net pyrradiometer was based on the same principles (Albrecht,
1933b), but both top and bottom plates were blackened. If T_1 and T_2
are the temperatures of the top and bottom plates, respectively, then the

net energy flux is given by the relation

$$F = (T_2 - T_1) [2\gamma + f(v)] \tag{11.2}$$

where γ is the thermal conductivity of the material between the plates and $f(v)$ is an empirical function of wind speed to be determined by measurements. A modified form of the Albrecht design was developed by Hofmann (1952), and the method of measurement by means of two blackened and thermally insulated plates was adopted for several of the modern net pyrradiometers.

11.3 IMPORTANT MODERN RADIOMETERS

11.3.1 Instruments with Unshielded Sensors

Any radiation sensor exchanges radiation with its environment, and the exchange takes place most freely if there is no window or other material surrounding the sensor. A number of different pyrgeometers, pyrradiometers, and net pyrradiometers have been designed to take advantage of such an unhibited exchange. Unfortunately, other types of energy transfers take place from an exposed sensor, principally conduction and convection effects from the ambient air, and some means must be used to account for nonradiative effects on the sensor. The three main ways of doing this are (1) to utilize a pair of sensors, one being shielded from radiation but not from convection and the other being exposed to both radiation and convection; (2) to ventilate the sensor with a constant jet of air which is strong enough to dominate the airflow over the sensor; (3) to correct the data themselves for wind effects on the sensor. The problems associated with the various methods are mentioned in the following discussions of individual instruments.

Courvoisier Net Pyrradiometer Although the net pyrradiometer developed by Courvoisier (1950) is not now generally available, it has played an important role as a comparison standard for the evaluation of other instruments (Moller, 1957). It is not designed for exposure to all types of weather, and is thus not suitable for routine recording of the radiation balance.

The device consists of two metal disks, one of which is exposed to the upward hemisphere and the other to the downward hemisphere, with an insulating layer between. The external surfaces of the disks are coated with Parsons Optical Black lacquer. Since this material has a higher absorptivity in the visible and near-infrared range than in the terrestrial range of wavelengths, the response of the instrument is somewhat selective. A jet of air is blown over the receiving surfaces to minimize wind effects in the measurements, the air jet being produced by an electric fan. The slots by which the airflow is controlled are not adjustable. A thorough test of the instrument (Ross and Sulev, 1962; Kirillova *et al.*, 1964) showed that in spite of the airflow over the sensing surfaces, the readings depend on the speed and direction of the prevailing wind, and that short-period fluctuations of up to 10% of the signal may be produced by atmospheric turbulence. A large and systematic error in the measurements is caused by the temperature of the air in the jet being higher than that of the surroundings. The elevation in temperature is caused by heat added by the blower and by the air chamber being heated by solar radiation. In cases of bright sunshine and little wind, the temperature error was as much as 5°C and caused the readings to be 2 to 3 times too high. An added difficulty is the fact that the temperature of the air in the channel ventilating the cold junctions of the thermopiles is up to 1°C higher than that ventilating the surfaces of the plates. This causes a significant and systematic error in the readings. One of the main conclusions from the tests was that the Courvoisier instrument is suitable for radiation measurements in the open air only at night, and even then care must be taken to account for wind effects.

Geir and Dunkle Pyrradiometer and Net Pyrradiometer   These instruments operate on the principle of the heat-flow plate (Geir and Dunkle, 1951). In essence, the sensor consists of two flat plates with a spacer between. If the plates are blackened and placed in a horizontal orientation in a radiation field, the temperature of the two surfaces will in general be different, the magnitude of the difference being a function of the relative fluxes of radiation in the upward and downward directions. In its original configuration, the sensor assembly consisted of a stack of three bakelite plates of dimensions $\frac{1}{64}$ in. thick by $4\frac{1}{2}$ in. square. The surfaces of the outside plates were covered with Fuller Flat Black Decoret paint. (The reflectance of this paint varies considerably with wavelength, thereby constituting a source of error which was never properly evaluated in the original models.) The center plate served as an insulator and as a mount for a silver-constantan thermopile, the junctions of which were in good thermal contact with two outside plates. The voltage V developed by the thermopile could be interpreted in terms of the net flux $F^{\uparrow} - F^{\downarrow}$ by the relation

$$F^{\uparrow} - F^{\downarrow} = kV \qquad (11.3)$$

where k is the calibration constant of the device. A small blower was used to maintain an airflow across the top and bottom surfaces to minimize wind effects and prevent the deposition of dew on the sensor. Wind tunnel tests showed a maximum error due to ambient wind of 4% for speeds up to 7 m/sec. By orienting the instrument so the jet from the blower was in the same direction as the wind, the error was reduced to 2% for a 13-m/sec wind. Short-period fluctuations were damped out by mounting aluminum plates between the thermopile and sensor plates, thus increasing the thermal capacity without much increase in thermal resistance. The time response was 12 sec for a 95% adjustment to steady state.

The pyrradiometer was of essentially the same construction as the net pyrradiometer except that a polished metal sheet was superimposed over the bottom plate to protect the plate from radiation from the lower hemisphere. The radiative flux on the top surface was then

$$F^\downarrow = kV + \sigma T^4 \tag{11.4}$$

where σ is the Stefan–Boltzmann constant and T is temperature of the plate in °K.

A commercial version* of the Geir and Dunkle net pyrradiometer is shown in Fig. 11.2. The sensing plates are painted with Parson's Optical

Fig. 11.2 Net pyrradiometer of Gier and Dunkle design as manufactured until recently by Beckman and Whitley and at present by Teledyne Geotech (photograph courtesy of Teledyne Geotech).

* The manufacturer was, until recently, Beckman and Whitley, San Carlos, Calif. Present manufacturer is Teledyne Geotech, 3401 Shiloh Road, Garland, Tex. 75040.

Black lacquer in this model. No correction is normally made for the wavelength selectivity of the painted surfaces, although measurements by various authors have shown that the absorptivity of this type of lacquer is about 5% higher in the visible and near-infrared wavelength ranges than at wavelengths beyond 5 μm. Thus a significant error is introduced due to wavelength selectivity unless the spectral distribution of the calibration source is the same as that of the radiation to be measured. In addition, the cosine response is also a function of wavelength, the longer wavelengths of terrestrial radiation being reflected more strongly at large angles of incidence than short-wave terrestrial radiation. For instance, at an angle of incidence of 80°, the reflectance of the sensor surface is about 7% at $\lambda = 10$ μm, 27% at $\lambda = 20$ μm, and 32% at $\lambda = 30$ μm.

The uncompensated models of the instrument are sensitive to temperature of the sensor assembly, thereby requiring a special measurement of this temperature by means of the copper-constantan thermocouple embedded in sensor. A reference junction must be provided by the user. The value of the calibration constant k is given at any temperature T by the relation

$$k_T = k_{26.7}[1 - (T - 26.7)/(T + 357)] \tag{11.5}$$

where $k_{26.7}$ is the value at the calibration temperature of 26.7°C. The correction $k_T/k_{26.7}$ varies from 1.19 at -40°C to 0.95 at $+50$°C. More recent models have a temperature compensating circuit built into the sensor, thereby obviating the auxiliary temperature measurement. The manufacturer claims for these models a temperature error of not more than ±1%. Another model has a thermistor network powered by batteries for generating this blackbody term of the sensor emission. The temperature error of this latter model is said to be not more than ±3% (Operation and Maintenance Manual, M-188, Teledyne Geotech, Oct. 23, 1972).

In addition to overall temperature effects and weathering of the exposed surfaces, the main sources of error in the Gier and Dunkle type instruments are the relatively large solid angle subtended at the sensing surfaces by the mounts of the sensors and by the case of the blower motor, the latter being particularly large for the lower sensor (see Fig. 11.2), wind effects for wind speeds more than about 5 m/sec^{-1}, and a nonuniform air flow over the surfaces. The latter two problems are somewhat alleviated by mounting the instrument on a mount which acts as a wind vane so that the air from the blower motor is always parallel to the ambient wind.

Yanishevsky Net Pyrradiometer   This is the principal instrument used for net radiation measurements in the climatological stations of the Soviet

Union. A general view of the instrument, with its shading device and cover, is shown in Fig. 11.3a. The sensing surfaces are 4.5 × 4.5 cm in dimensions, and are made of blackened copper foil. They are mounted on a disk which is 10 cm in diameter and 1.5 mm thick, which is itself attached to a handle on a ball joint pivot.

The body of the sensor consists of the following layers, listed in order from top to bottom (Yanishevsky, 1957; Reifer *et al.*, 1971): blackened copper foil 0.04 mm thick; insulating paper 0.02 mm thick; series of silver-constantan thermojunctions, insulating paper; side-by-side copper bars 2.5 mm thick by 5.5 mm wide; insulating paper; second series of thermojunctions; insulating paper; blackened copper foil. The thermopile consists of 10 sections of about 60 thermojunctions each, one section on each of the 10 copper bars as shown in Fig. 11.3b. Each section is made by winding 32 turns of constantan ribbon 0.85 mm wide and 0.03 mm thick around the insulated copper bar, and plating one-half of each turn with silver to a thickness of 0.01–0.04 mm. The ten sections are connected in series, thereby making a thermopile of something over 600 junctions which has a resistance of about 40 Ω and a sensitivity of 7 to 8 mV/cal cm^{-2} min^{-1}. The time constant of the instrument is about 12 sec (presumably for full response). It operates satisfactorily throughout the temperature range −60° to +60°C. When not in use, the sensor is protected with the cover provided; the

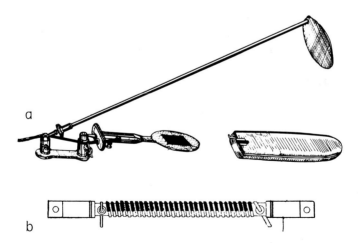

Fig. 11.3 (a) Configuration of the Yanishevsky net pyrradiometer, with shading device and protective cover. (b) Schematic diagram of one of the ten copper bars with its silver-constantan thermopile (adapted from Kondratyev, 1969; and Yanishevsky, 1957).

shading device is used for calibration by reference to the measurements from a pyrheliometer.

Since the sensor has neither artificial ventilation nor protective window, the output of the device is strongly affected by the wind. In order to take account of this source of error, each instrument is individually calibrated for wind effects, the wind is measured at the same time as the radiation, and corrections are applied during data processing. Typical values of the correction factors for various wind speeds, as given by Yanishevsky (1957), are tabulated below.

Wind speed, m sec⁻¹	1	2	3	4	5	6	7	8	10	12	16	60
Correction factor	1.03	1.05	1.08	1.1	1.13	1.15	1.17	1.2	1.22	1.25	1.31	2.0

Data on the wavelength dependence of the absorptivity of the sensing surfaces is not immediately available. Ross and Sulev (1962) found that the sensitivity of one of the instruments was 25 to 35% lower for long-wave radiation than for solar radiation, but this result needs further confirmation. A general survey of the problem is given by Ross et al., (1964).

Net Pyrradiometer of Suomi, Franssila, and Islitzer   This instrument is based on the Geir and Dunkle design, but some of the problems are minimized in this device. The details of sensor construction and the adjustable vane for controlling air flow over the sensors can be seen in Fig. 11.4 (Suomi et al., 1954). The sensor is constructed by winding 120 turns of 0.2-mm-diameter constantan wire around a microscope slide of dimensions 0.11 × 2.5 × 7.5 cm and plating half of each turn with copper. A heating coil is made by interlaying insulated copper wire between the thermoelements. The sensing surface is made by covering the exposed wires with Fuller's Velvet-Black Decoret enamel.

The most important advantages of this net pyrradiometer over that of Teledyne Geotech are the provision of an adjustable vane (see Fig. 11.3) for controlling the air flow over the two sensor surfaces and an electric heater on the radiation plate to furnish a sensitive control of the ventilation. In addition, 6% of the thermopile junctions are covered with white paint, which has a high reflectivity in the solar radiation range and low reflectivity

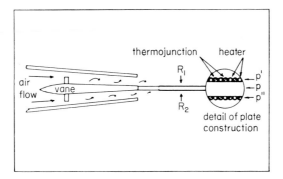

Fig. 11.4 Schematic diagram of adjustable vane for controlling ventilation and constructional details of the sensor of the net pyrradiometer of Suomi, Franssila, and Islitzer (after Suomi *et al.*, 1954).

in the terrestrial range, to decrease the errors due to wavelength selectivity shown by Fuller's Decoret enamel. Performance characteristics for the device, as given by Suomi *et al.* (1954), are shown in the following tabulation.

Instrument sensitivity: 1.3 mV/cal cm^{-2} min^{-1}
Wind effects:
    wind 6 m/sec: no effect
    wind 12 m/sec with jet directed downstream: no effect
    wind 12 m/sec with jet directed upstream: $+ 0.3\%$
    crosswind of 18–12 m/sec^{-1}: $+ 2\%$
Cosine law: decrease of 1% for direction of incidence varying
    from $0°$ to $90°$
Speed of response: 100% response in 5 sec
Indicated overall accuracy: net radiation to $\pm 2\%$

Net Pyrradiometer of the Kew Observatory   This instrument is of the basic design described by McDowall (1955). As can be seen by the photo in Fig. 11.5, it is a ventilated instrument similar to that of Geir and Dunkle (1951). The sensor is a pair of thin aluminum plates 7.6 cm square and blackened with an optical black paint developed at the National Physical Laboratory. The sensor plates are separated by a bakelite frame, which is slotted to hold the elements of the 120-junction copper-constantan thermopile, and two thin sheets of polyethylene used for electrical insulation. A unique feature of this design is the extension of air ducts along the sides of the sensor plate in order to decrease the effects of a crosswind.

Fig. 11.5 Net pyrradiometer of the Kew Observatory (photograph by author).

Testing of instrument operation showed a moderate sensitivity to wind speed and direction. A 5 m sec^{-1} crosswind decreases the output by about 1%, and a 10-m sec^{-1} crosswind by about 3%. The effect for a wind opposing the radiometer stream is somewhat complicated, but errors of ±1% occur over the speed range of 0–10 m sec^{-1}. Shielding of the sensor elements by the structural members results in an effective decrease of aperture by 3% for isotropic radiation, and the cosine response was found to be "satisfactory." The calibration constant was found to vary within the range ± 5% over a 6-months period of operation.

Other Radiometers with Unshielded Sensors   A number of other radiometers with unshielded sensors have been devised, but none appears to be in general use at the present time. One of the first was the net pyrradiometer of Falkenberg (1947), which consisted of a 24-junction manganin constantan thermopile mounted at the backs of two blackened sheets of mica. A unique feature of this device was that the entire sensor was rapidly vibrated by means of an electric motor in order to minimize the effects of wind. The method worked successfully for wind speeds up to about 10 m sec^{-1}.

The Wagner net pyrradiometer was designed on the compensation principle (Frankinberger, 1954), and without ventilation. The top and bottom sensor plates were heated to the same temperature in order to make the heat transfers between the plates and the surrounding air the same, and thus eliminate wind effects. The net radiation was then given as the difference in the amount of heat supplied to the two surfaces. The assumption of equal heat transfer from top and bottom surfaces was not realized in practice, particularly in cases of well-developed turbulence due to surface heating, and was a source of large errors in the measurements. The instrument was best suited to obtaining daily or monthly totals of net radiation.

The Laikhtman–Koocherov pyrradiometer, as described by Kondratyev (1965), consisted of a blackened receiving disk of copper, shielded from below and exposed to radiation from the upper hemisphere. In operation, the disk was cooled below ambient air temperature and then permitted to heat up by radiation and conduction from the air. The rate of rise of temperature of the plate at the instant its temperature exactly equalled that of the surrounding air could be interpreted in terms of the radiative flux incident on the receiver plate.

11.3.2 Instruments with Shielded Sensors

The most successful method of minimizing or eliminating wind effects and weathering in measurements of radiation is by shielding the sensors with some type of window material which transmits the radiation freely. Although no material has been found which is completely transparent to radiation of all wavelengths, materials such as polyethylene and KRS-5 provide a reasonable approximation to complete transparency. As described below, these and certain others have been used for shielding the sensors of different types of pyrradiometers and net pyrradiometers.

CSIRO Net Pyrradiometer   This instrument was developed by Funk (1959) of the Commonwealth Scientific and Industrial Research Organization (*CSIRO*) of Australia. Three different models of the instrument (the standard stationary model, a portable model, and a miniature model for biological applications) are now available commercially.* The sensor consists of two rows of 125 thermojunctions each, separated by a sheet of plastic and sandwiched between, but insulated from, two aluminum plates of approximate dimensions 3×4 cm. The plates are sprayed with Parsons Optical Black lacquer, and covered with hemispherical shields made of 0.05-mm-thick polyethylene. A ring with a low-voltage heating element surrounds the sensor head, the element being positioned so as to

---

* Sold by Middleton Instruments, 75–79 Crockford Street, Port Melbourne, Australia.

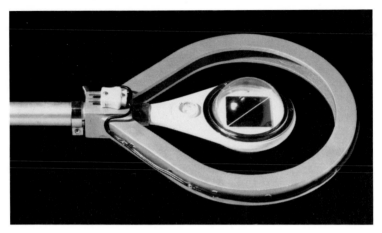

Fig. 11.6 Photograph of the CSIRO net pyrradiometer sensing head and heating ring (courtesy of Middleton Instruments).

prevent the formation of dew on the polyethylene domes without irradiating the sensor surfaces. A photograph of the sensor head and heating ring is shown in Fig. 11.6. In operation, the polyethylene hemispheres are kept inflated by a very slow stream of dry nitrogen introduced by a tube from a pressure reduction valve on an ordinary nitrogen gas cylinder. The flow is made sufficiently slow for one cylinder of gas to last for 6 months.

The sensitivity of the device is about 30 mV/cal cm^{-2} min^{-1}, with an internal resistance of 80 Ω. The spectral selectivity is minimized by covering about 5% of the sensor surfaces with a paint which is highly reflecting in the solar wavelength range but essentially black at long wavelengths. Wind has a negligible effect on instrument output. The deviation from a cosine response is not greater than $\pm$ 1.5% over a range of incident angles of 0° to 75°, according to the manufacturer's specifications. The instrument may be made into a pyrradiometer for making measurements of flux from a single hemisphere by covering one of the sensor surfaces with a polished metal adapter.

Other models of the CSIRO net pyrradiometer include a portable model, with the necessary accessories for transport on expeditions, and a miniaturized version, with a sensor head only 1.2 cm diameter, for use in biological applications. An interesting adaptation of the CSIRO net pyrradiometer for "solar-blind" operation was developed by Paltridge

(1969) by the use of a black polyethylene spherical shell surrounding the entire sensor head. The shell was kept inflated by a slow stream of nitrogen, and in order to counteract the effects of solar heating of the upper surface of the black polyethylene, the whole shell was rotated at 5 Hz by a small electric motor. The principal limitation of the device is that the wavelength dependence of transmission of radiation through the black polyethylene makes the measurements difficult to interpret.

Schulze Net Pyrradiometer Three different models of this instrument have been developed, the latest of which has been described in detail by Dake (1972). The core of the instrument is an aluminum cylinder of 10 cm diameter and 26 mm height which is machined to produce five cooling fins around the outside (see Fig. 11.7). This core is supported inside an exterior housing and covered on top and bottom by radiation shields of the type shown in the bottom part of Fig. 11.7. Air is forced by an electric blower into the instrument cavity and escapes through a combination of slits surrounding the polyethylene domes and escape holes in each of the top

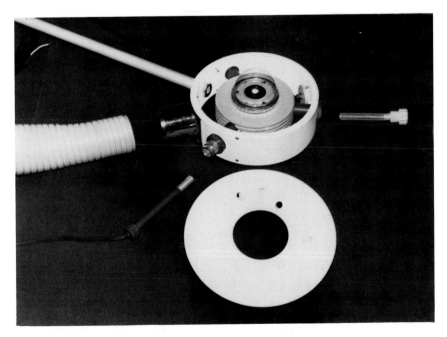

Fig. 11.7 Photograph of a partially assembled Schulze net pyrradiometer (courtesy of Dr. F. Kasten, Meteorologisches Observatorium, Hamburg).

and bottom radiation shields. The holes are bored on an off-center slant to cause the escaping air to circulate around the polyethylene shields for temperature control and prevention of dew on the shields. The circulation of air is strong enough to keep the entire instrument at ambient air temperature under normal operating conditions, but a heating element can be introduced into the air stream for use in hostile environments. The temperature of the aluminum core is monitored by means of a 6-element thermopile inserted as shown in Fig. 11.7. A small capsule of desiccant is also inserted into the core for preventing moisture condensation inside the instrument.

The sensors of the first two models of the Schulze net pyrradiometer are a pair of silver-constantan thermopiles with a sensitivity of 18–24 mV/cal cm^{-2} min^{-1} (Schulze, 1953). A unique feature of the latest model* are semiconductor thermopiles consisting of 48 sheets of Bi–Sn–Sb bismuth alloy connected in series ("Thermagotrons" of MCP Electronics, Alperton, England), with a sensitivity of 65 mV/cal cm^{-2} min^{-1} and an internal resistance of 3 Ω. One undesirable feature of these thermopiles is a relatively large temperature sensitivity. Data of Dake (1972) show a variation of + 4 to −13% over the temperature range −10 to + 40°C. No temperature compensation is provided, but appropriate corrections can be made on the basis of the measured instrument temperature. The time constant for 99% adjustment of the signal is 150 sec, and the wavelength selectivity of the instrument is minimized by covering the center portion of the blackened receiver with white paint.

Although there are many Schulze net pyrradiometers in use, and apparently giving satisfactory measurements, tests by Ross and Sulev (1962) of some of the older models indicated significant difficulties with them. For instance, the sensitivities of two back to back instruments constituting the net pyrradiometer differed by 20–40%, thereby making the instrument unsuitable for net radiation measurements, the sensitivity was 16–20% lower for long-wave than for short-wave radiation, the polyethylene shields were nonuniform in optical characteristics, and in spite of the enclosed desiccant, water condensed on the inside of the hemispheric shields. Presumably these effects have been minimized or eliminated in the later models.

Fritschen Miniature Net Pyrradiometer The basic design of this instrument is similar to that of the CSIRO net pyrradiometer, but the sensor head is only about 6 cm in diameter. The original model (Fritschen, 1963) incorporated hemispheric shields of polystyrene, but because of poor transmission properties of this material, it was replaced by polyethylene (Fritschen,

---

* Brochure from the manufacturer, Dr. Bruno Lange, 14–18 Hermannstrasse, Berlin (1973).

1965). The transducer of the latest version is a 22-junction thermopile of manganin-constantan with 38-Ω internal resistance which yields a
sensitivity of about 3.5 mV/cal cm^{-2} min^{-1} for the temperature compensated model and 4.2 mV/cal cm^{-2} min^{-1} for the uncompensated model.
The temperature coefficient of the first of these is 0.23%/°C over the
30–70°C range, while that of the second is only 0.07%/°C. The hemispheres
are kept inflated and internal condensation of moisture is prevented by a
continual pressurization or periodic inflation with dry air. A 10% difference
in sensitivity between short-wave and long-wave radiation is compensated
for by 10% of the receiving surfaces being covered with white paint. A
modification of the Fritschen instrument by the use of thermistors to
provide larger signals was suggested by Campbell *et al.* (1964).

Thornthwaite Miniature Net Pyrradiometer   This small net pyrradiometer,
the sensor head of which is 5 cm in diameter, is useful in biological applications. The blackened sensor surfaces are shielded by polyethylene
hemispheres which are kept inflated with dry air through two purge
orifices. A photograph of the instrument is shown in Fig. 11.8. Two models

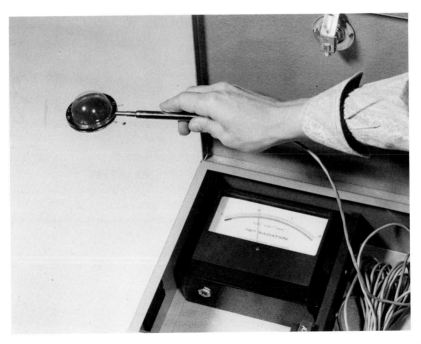

Fig. 11.8 Photograph of the Thornthwaite miniature net radiometer (courtesy of
C. W. Thornthwaite Associates).

of the device, with sensitivities of 0.25 and 3.5 mV/cal cm⁻² min⁻¹, are available. According to the manufacturer,* the response follows the cosine law with angle of incidence, and is "virtually independent" of ambient temperature and wind.

Pyrradiometer and Net Pyrradiometer of the Physico-Meteorological Observatory, Davos This instrument utilizes the Moll thermopile with 60 copper-constantan thermoelements arranged to form a 25-mm-diameter circular sensing surface. The hot junctions of the thermopile are in good thermal contact with the blackened receiver, and the cold junctions are in thermal contact with the aluminum case. Two different types of thermopiles, yielding two different sensitivities, are available. Typical specifications supplied by the manufacturer† on the two types are given in the following tabulation.

Type of thermopile	TD 25-60C	TD 25-60
Sensitivity (mV/cal cm⁻² min⁻¹)	7	22
Time constant: $1/e$ (sec)	2	4
Internal resistance (Ω)	170	180
Temperature coefficient (% per °C)	−0.1 (max)	−0.15 (max)
Cosine error at 80° angle of incidence (%)	−3.0 (max)	−3.0 (max)

These basic sensors have been used for both pyranometers and pyrradiometers, the main difference between the two being the type of hemispheric shield mounted over the sensor. Optical glass is used for the pyranometer and polyethylene (Lupolen) of 0.05 mm thickness for the pyrradiometer. A temperature sensor, either a thermoelement or a platinum resistance thermometer, for measuring the temperature of the receiver is built into the pyrradiometer. The net pyrradiometer is made by mounting two pyrradiometers in a back-to-back configuration, the net radiation being computed from the difference between the two signals. An installation consisting of two pyrradiometers (left-hand side of diagram) and two pyranometers mounted to form a net pyrradiometer and an albedometer is shown in Fig. 11.9.

Eppley Pyrradiometer This instrument, designed to measure the longwave flux from one hemisphere, has some novel features, including thermoelectric temperature control of the sensing surface to any selected temperature, and automatic correction of the output signal for radiation emitted by

* Thornthwaite Associates, Elmar, N. J. 08318.

† The manufacture of these instruments by the Physico-Meteorological Observatory, CH 7270, Davos-Platz, Switzerland was suspended in 1971. According to Dr. C. Frolich in June, 1973, a search for an industrial manufacturer was under way.

Fig. 11.9 Two pyrradiometers of the Physico-Meteorological Observatory, Davos mounted back-to-back to form a net pyrradiometer (left half of photograph) and two pyranometers mounted as an albedometer (courtesy of the Physico-Meteorological Observatory).

the surface (Drummond *et al.*, 1968). Thus the output is dependent on only the incident radiant flux and not on the difference between incident and emitted fluxes, as in most pyrradiometers. The sensor is a 100-junction copper-constantan thermopile which is covered with a 19-mm-diameter plastic film and painted with Parsons Optical Black lacquer. The 50-mm-diameter hemispherical shield over the sensor is made of KRS-5, on the inside of which is deposited several layers of material which constitute an interference filter for the isolation of energy in the 4–50 μm wavelength range. Thus the instrument is in the "solar blind" category.

Specifications for the instrument, as given by the manufacturer* are given in the following tabulation.

Sensitivity	5 mV/cal cm^{-2} min^{-1}
Impedance	400 Ω
Temperature dependence	$\pm$ 0.5% over range $-$ 20 to $+40°$C
Linearity	$\pm$ 1% over range 0–1 cal cm^{-2} min^{-1}
Response time $(1/e)$	2 sec
Cosine response	better than 5% from normalization

* Eppley Laboratory, Newport, R.I. 02840.

In appearance, the instrument is similar to the Eppley precision pyranometer (see Chapter 4). The case of the developmental unit was of cast aluminum with heat transfer fins around the outside (Drummond *et al.*, 1968), but this has been supplanted in the commercial model by a chrome-plated case, in which a replaceable desiccant is provided. Temperature of the sensing element is monitored continuously by a thermistor, which controls the battery voltage supplied to the temperature control circuit.

Georgi Universal Radiation Meter   This device was originally designed by Georgi (1956) as an instrument which could be used for many different types of radiation measurements: global, direct, and diffuse solar radiation, albedo, total radiative flux (either upward or downward) of combined solar and terrestrial radiation, and broad-band spectral distributions. The sensor is a blackened Moll–Gorczynski-type thermopile (see description in Chapter 4) with a sensitivity of about 7 mV/cal cm^{-2} min^{-1}, and 7 Ω internal resistance.* Interchangeable covers of glass and polyethylene are provided for solar or terrestrial radiation measurements. The collimator tube used for direct solar radiation measurements is fitted with various types of Schott broad-band absorption filters. The temperature of the sensor may be measured with a mercury thermometer, which is inserted into the thermopile holder through a special orifice, for applying temperature corrections to the measurements. The signal is read by means of a dual-range microammeter.

Pyrradiometer of Stern and Schwartzmann   The sensing element of this instrument (Stern and Schwartzmann, 1954) was a chromel-constantan thermocouple welded to a nickel foil disk 6 mm in diameter. The disk was covered with gold–black and mounted near the optical center of a KRS-5 (thallium bromoiodide) hemisphere of 2.7 cm inside diameter. The temperature of the housing for the unit was made equal, to within ± 0.1°C, to that of the ambient air by a rapid circulation of a liquid from the coils of an air conditioner unit to the double wall of the instrument housing.

Tests of the device showed the output to be independent of wind speed over the range 0–5.4 m/sec, and the time constant for 100% response was only 0.2–0.3 sec. The single thermocouple sensor resulted in the very low sensitivity of about 0.44 mV/cal cm^{-2} min^{-1}. The angular response showed errors of about 5 and 12% for incident angles of 30° and 60°, respectively. No evidence has been found that more than one of the devices was ever produced.

---

* Specifications from Kahl Scientific Instrument Corp., P.O. Box 1166, El Cajon, Calif. 92022.

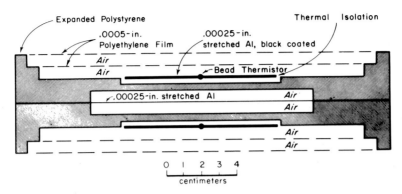

Fig. 11.10 Cross section of the latest model of the Suomi–Kuhn economical net pyrradiometer (after Kuhn, 1970).

Suomi-Kuhn Economical Net Pyrradiometer Economy is the watchword of this instrument (Suomi and Kuhn, 1958), several models of which have been developed.* It is sufficiently simple and light in weight (about 90 g) for attachment to an ordinary radiosonde for measurements of the vertical profile of net radiation, and indeed this has been its principal use (Suomi *et al.*, 1958; Kuhn, 1963, 1970). The main frame of the instrument consists of two approximately circular pieces of expanded polystyrene about 3 cm thick and 21 cm in diameter. A cross-sectional diagram of the latest model, as given by Kuhn (1970) is shown in Fig. 11.10. The sensor surface is a blackened aluminum foil 0.006 mm thick and 9 cm in diameter, the temperature of which is measured by an attached bead thermistor. Heat conduction between sensor and mount is minimized by supporting the aluminum foil with eight fine threads of low heat conductivity. An air space bisected by another aluminum foil is used to minimize heat transfer between the top and bottom sensors, and a double layer of stretched polyethylene film shields each of the sensors from the effects of wind.

The overall accuracy for measurements of net radiation with the device has apparently not been determined precisely, but Suomi and Kuhn (1958) obtained results with this instrument similar to those of the ventilated radiometer of Suomi *et al.* (1954) if the readings were averaged over 1 hr. The flat polyethylene shields produce a poor cosine response of the instrument. For instance, for an angle of incidence of $\theta_0 = 25°$, the ratio of readings between the ventilated and economical radiometers was 1.1; at $\theta_0 = 85°$ it was 1.37. The response time of the instrument for a signal adjustment of $1/e$ was given by Kuhn (1970) as 25 sec for the improved model, whereas it was about twice that in the original design.

* Manufactured by UNECO, Inc., P.O. Box 487, Bellevue, Neb. 68005.

The device has been manufactured in different configurations, including one triangular model. It has been used most frequently as an attachment to a radiosonde for measuring vertical profiles of net radiation, but its low cost makes it adaptable to the establishment of concentrated networks for special radiation studies.

11.4 CALIBRATION OF PYRRADIOMETERS AND NET PYRRADIOMETERS

Because of weathering of receiving surfaces, aging of materials, electronic drifts, and other factors, long-wave radiometers in routine use must be calibrated frquently in order to maintain reasonable accuracy in the measurements. A calibration once every 6 months is desirable for any of these devices, and those with unshielded sensors should be calibrated more frequently than this.

There are three different methods in normal use for calibrating hemispherical infrared receivers, the sun-and-shade method, the method of using a transfer standard instrument for which the calibration factor is already established, and the method using a special infrared calibrator. The first two of these are simple, convenient, and suitable for field application, while the latter provides a higher calibration accuracy, particularly for the long-wave part of the spectrum. A fourth, and also relatively simple, method of calibration has recently been proposed by Idso (1971). Each of these will be briefly outlined below.

In order to use the Sun-and-shade calibration method, it is necessary to use an auxiliary pyrheliometer, which is itself well calibrated, and a disk fixed to a long handle for shading the sensor surface from the direct rays of the Sun. The pyrheliometer and radiometer to be calibrated are set up alongside each other near noon on a day with little cloudiness and good optical stability of the atmosphere. The flux of direct solar radiation F_0 is measured at the same time the sensor of the long-wave radiometer is shaded from the direct sunlight. If V_1 is the output signal of the radiometer just before shading and V_2 is the signal during shading, then the calibration constant k of the instrument is given by

$$k = (V_1 - V_2)/F_0 \cos \theta_0, \qquad (11.6)$$

θ_0 being the zenith angle of the Sun. The units of k in many of the commercially available instruments are mV/cal cm^{-2} min^{-1}, although this is not universally the case.

In carrying out the Sun-and-shade calibration the shading disk should be held at least $1\frac{1}{2}$ to 2 m from the receiving surface and should be of a size such as to subtend approximately the same solid angle at the receiving surfaces as that subtended by the pyrheliometer. The latter varies among the different pyrheliometers (see Chapter 3), but a 5° total angle cone is a good approximation for the purpose. Care must be taken to assure that the shading of the surface is complete, including both the umbra and penumbra portions of the shadow. In case a pyrheliometer is not available, a well-calibrated pyranometer can be substituted, in which case $F_0 \cos \theta_0$ is obtained by the Sun-and-shade method for the pyranometer also.

The accuracy of the calibration of a long-wave radiometer by the Sun-and-shade method is dependent on the accuracy of the comparison instrument, and on the extent to which the response of the long-wave device is independent of wavelength. All of the blackening materials found so far for covering the receivers do have a wavelength selectivity, and the ruse of painting 5 or 6% of the receiver surface with a white paint to eliminate the selectivity is only partially successful. In fact, Idso (1970) has obtained evidence that no white paint should be applied for polyethylene shielded radiometers. In spite of these difficulties, the Sun-and-shade method used properly gives at least consistent data on long-wave fluxes, although the absolute accuracy obtainable is still an open question.

The calibration of one radiometer by means of another radiometer, a transfer standard, for which the calibration factor is already well established is so simple as to need little explanation. The two are set up side-by-side over a uniform surface, and the value of the net or hemispherical radiative flux measured by the transfer standard is used to compute the unknown calibration factor of the instrument being calibrated. The main requirement is that the standard instrument be a high-quality device which has a stable response and for which temperature, wind, and other environmental effects are adequately known and accounted for. Ideally, the standard will have been calibrated by the calibration chamber method described below, and will be carefully handled and used only as a transfer standard. The separate responses of the operational instrument to solar and terrestrial radiation can be determined separately by the Sun-and-shade method in the daytime and by comparison with the transfer standard at night.

The most accurate calibration of long-wave radiometers is obtained by means of a special calibration chamber, such as that described by Mac-Dowall (1955) and used in modified form by Funk (1959) and Fritschen (1963). Basically, the sensor element of the instrument is placed inside a blackened chamber and positioned near an aperture cut into the chamber wall. A blackbody radiation source (blackbody cavity) is then positioned at the aperture to furnish radiation for the calibration. The temperatures T_1, T_2, and T_3 of the cavity, the chamber, and the sensing element, respectively, are measured, and that of the cavity, T_3, is variable and closely controlled. The radiative energy flux received by the surface facing the cavity is (Funk, 1959)

$$F_1 = \psi\sigma T_1{}^4 + \epsilon(1 - \psi)\sigma T_2{}^4 - \sigma T_3{}^4 + (1 - \epsilon)(1 - \psi)\sigma T_3{}^4 \quad (11.7)$$

where ϵ is the emissivity of the walls of the chamber, σ is the Stefan–Boltzmann constant, and ψ is the effective aperture. ψ is given by the ratio $\psi = J/J'$, where J' is the blackbody radiation the element would receive from a hemispheric blackbody and J is that actually received from the cavity source. The first two terms on the right-hand side of Eq. (11.7) are the fluxes received from the cavity and from the walls of the chamber, respectively, the third is that emitted by the element, and the last is that due to reflection from the walls of the chamber. The flux on the back surface is similarly

$$F_2 = \epsilon\sigma T_2{}^4 - \sigma T_3{}^4 + (1 - \epsilon)\sigma T_3{}^4 \quad (11.8)$$

and the net flux is

$$F_1 - F_2 = \psi\sigma[T_1{}^4 - \epsilon T_2{}^4 - (1 - \epsilon)T_3{}^4] \quad (11.9)$$

The accuracy of this method of calibration is mainly dependent on the accuracy to which the emissivity of the chamber walls is known, the accuracy of the various temperature measurements, and the degree to which nonradiative heat transfers can be avoided. Careful attention to detail should yield values of the calibration constant certainly to $\pm$ 2%, and perhaps better than that, which becomes comparable to other errors of measurement.

A third method of calibrating pyrradiometers and net pyrradiometers, suggested by Idso (1971), utilizes a large temperature-controlled flat black plate in a laboratory setting. The sensing element is positioned directly over, or under, the plate. If the room conditions are constant, the net radiation F_n sensed by the instrument is

$$F_n = F_c - \psi\epsilon\sigma T^4 \quad (11.10)$$

where F_c is the constant net radiation due to the surroundings, ϵ and T are emissivity and temperature of the plate, respectively, σ is the Stefan–Boltzmann constant, and ψ is the "angle factor" for radiative exchange between plate and sensor. A nomogram for determining ψ for different conditions is given by Idso (1971). The output signal V of the instrument is

$$V = cF_c - k\psi\epsilon\sigma T^4 \quad (11.11)$$

where c is the (unknown) calibration factor for the combined short- and long-wave radiation of the room, and k is the calibration factor for long-wave radiation. For the calibration, values of V are obtained while the temperature of the plate is varied. Then the value of k can be determined by the slope of a curve of V plotted versus the quantity $\psi \epsilon \sigma T^4$.

From an analysis of the probable errors of the various factors in Eq. (11.11), Idso concluded that the accuracy obtainable by the method is $\pm 3\%$. This value was further substantiated by comparisons with factors obtained by use of a calibration chamber.

REFERENCES

Abbot, C. G., Fowle, F. E., and Aldrich, L. B. (1922). *Ann. Astrophys. Obs. Smithsonian Inst.* 4.

Albrecht, F. (1933a). Apparate und messmethoden der atmospharischen strahlungsforschung. *In* F. Linke, "Meteorologischen Taschenbuch," Vol. II, 46–109. Akademische Verlagesellschaft, Leipzig.

Albrecht, F. (1933b). Ein strahlungsbilanzmesser zur messung der strahlungshaushaltes von oberflachen. *Meteorol. Zeit.* **50**, 62–65.

Ångström, K. (1905). Ueber die anwendung der elektrischen kompensationsmethode zur bestimmung der nachtlichen ausstrahlung. *Nova Acta Soc. Sci. Upsala, Ser. 4* **1**, No. 2.

Budyko, M. E. (1972). Private communication.

Callendar, H. L. (1910). The radio-balance. A thermoelectric balance for the absolute measurement of radiation, with applications to radium and its emanation. *Proc. Phys. Soc. London, Sect. A* **23**, 1–34.

Campbell, G. S., Ashcroft, G. L., and Taylor, S. A. (1964). Thermistor sensor for the miniature net radiometer. *J. Appl. Meteorol.* **3**, 640–642.

Courvoisier, P. (1950). Uber einen neuen strahlungsbilanzmesser. *Verh. Schweiz. Naturforsch. Ges.* **130**, 152.

CSAGI (1958). Radiation instruments and measurements. *Ann. Int. Geophys. Yearbk.* **5** (Part 4), 367–466.

Dake, C. U. (1972). Uber ein neues modell des strahlungsbilanzmessers nach Schulze. *Deutscher Wetterdienst, Berichte* **16**(126), 1–22.

Drummond, A. J., Scholes, W. J., Brown, J. H., and Nelson, R. E. (1968). A new approach to the measurement of terrestrial long–wave radiation, W.M.O. Tech. Note No. 104, Proc. W.M.O./I.U.G.G. Symp., Bergen, Norway, Aug. 1968.

Falkenberg, G. (1947). Ein vibrationspyranometer. *Meteorol. Z. (Potsdam)* **1**, 372.

Frankenberger, W. (1954). Ergebnisse von warmehaushaltsmessungen. *Meteorol. Rundsch.* **7**, 81–85.

Fritschen, L. J. (1963). Construction and evaluation of a miniature net radiometer. *J. Appl. Meteorol.* **2**, 165–172.

Fritschen, L. J. (1965). Miniature net radiometer improvements. *J. Appl. Meteorol.* **4**, 528–532.

Funk, J. P. (1959). Improved polyethylene-shielded net radiometer. *J. Sci. Instrum.* **36**, 267–270.

Geir, J. T., and Dunkle, R. V. (1951). Total hemispherical radiometers. *Trans. Amer. Inst. Elec. Eng.* **70**, 339–343.

Georgi, J. (1956). Meteorologischer universal-strahlungsmesser. *Meteorol. Rundsch.* **9**, 89–92.

Hofmann, G. (1952). Ein strahlungsbelanzmesser fur forstmeteorologische untersuchungen. *Forstwissenschaft. Centralblatt* **71**, 330–337.

Idso, S. B. (1970). The relative sensitivities of polyethylene shielded net radiometers for short and long wave radiation. *Rev. Sci. Instrum.* **41**, 939–943.

Idso, S. B. (1971). A simple technique for the calibration of long-wave radiation probes. *Agr. Meteorol.* **8**, 235–243.

Kirillova, T. V., Ross, Yu. K., and Sulev, M. A. (1964). Comparison of balance meters. W.M.O./I.U.G.G. Rad. Symp., Leningrad, U.S.S.R. (Aug., 1964).

Kondratyev, K. Ya. (1965). "Actinometry." Hydrometeorological Publishing House, Leningrad, U.S.S.R.

Kondratyev, K. Ya. (1969). "Radiation in the Atmosphere." Academic Press, New York.

Kuhn, P. M. (1963). Sounding of observed and computed infrared flux. *J. Geophys. Res.* **68**, 1415–1420.

Kuhn, P. M. (1970). Applications of thermal radiation measurements in atmospheric science. *In* "Advances in Geophysics" (A. J. Drummond, ed.), Vol. 14. Academic Press, New York.

MacDowall, J. (1955). Total-radiation fluxmeter. *Meteorol. Mag.* **84**, 65–71.

Moller, F. (1957). Radiation comparisons at Hamburg. *W. M. O. Bull.* **6**, (No. 1), 13.

Paltridge, G. W. (1969). A net long-wave radiometer. *Quart. J. Roy. Meteorol. Soc.* **95**, 635–638.

Reifer, A. B., Alekseenko, M. I., Buertsev, P. N., Zastenker, A. I., Beloguerov, Yu. A., Nepomnyashchii, S. I. (1971). "Collection on Hydrometeorological Instruments and Stations." Hydrometeorological Publishing House, Leningrad, U.S.S.R.

Robinson, G. D. (1964). Surface measurements of solar and terrestrial radiation during the IGY and IGC. *Ann. Int. Geophys. Yearbk.* **32**, 17–61.

Ross, Yu. K., and Sulev, M. (1962). Some results of the comparison of Courvoisier, Schulze, and Yanishevsky balancemeters. *Geofis. Pura Apl.* **53**, 88–100.

Ross, Yu. K., Sulev, M. A., Yanishevsky, Yu. D. (1964). Present status of measurements of the radiation balance and its longwave composition at the Earth's surface. Trudy, Mezhvedomstvennoe Soveshchanie po Aktinometrii i Optike Atmosfery, 5th, Moscow, pp. 10–24.

Schulze, R. (1953). Uber ein strahlungsmessgerat mit ultrarotdurchlassiger windschutzhaube am Meteorologischen Observatorium Hamburg. *Geofis. Pura Appl.* **24**, 107–114.

Stern, S. C., and Schwartzmann, F. (1954). An infrared detector for measurement of the back radiation from the sky. *J. Meteorol.* **11**, 121–129.

Suomi, V. E., and Kuhn, P. M. (1958). An economical net radiometer. *Tellus* **10**, 160–163.

Suomi, V. E., Franssila, M., Islitzer, N. F. (1954). An improved net radiation instrument. *J. Meteorol.* **11**, 276–282.

Suomi, V. E., Staley, D. O., and Kuhn, P. M. (1958). A direct measurement of infrared radiation divergence to 160 mb. *Quart. J. Roy. Meteorol. Soc.* **84**, 472–473.

Yanishevsky, Yu. D. (1957). "Actinometric Instruments and Methods of Observation." Hydrometeorological Publishing House, Leningrad, U.S.S.R. (in Russian).

List of Symbols

A area; absorptivity; albedo; a constant; indicator of aerosol; parameter for scattering matrix; matrix element in Curtis method of computation

B Planck function; a constant; parameter for scattering matrix; volume attenuation coefficient; luminous intensity

C a constant; specific heat; element of Mueller matrix; cooling rate; Planck radiation constants

D indicator for Doppler; reciprocal of filter transmission

D^* measure of merit of a radiation detector

E energy

F flux of radiation

G global (hemispheric) flux of solar radiation

H heating rate

I intensity; a Stokes parameter

J source function; Bessel function

K absorption coefficient; calibration constant; luminous efficiency

L luminous flux

M parameter in cooling-rate calculations

N number density of particles; relative sunspot number

O indicator for ozone

P degree of polarization; phase function, matrix, or matrix element; parameter in determination of Linke turbidity factor; Legendre polynomial; Mueller matrix

Q radiant flux on a horizontal surface; a Stokes parameter; luminous energy

R reflectance; distance or length; daily global flux of solar radiation; indicator for Rayleigh

S scattering function or matrix; element of Mueller matrix; intensity of absorption line; fraction of possible sunshine duration; ratio of Sun–Earth distance to its mean value

T temperature; Linke turbidity factor; transmission function or transmissivity; optical thickness

U a Stokes parameter

V a Stokes parameter; electromotive force

W indicator for water vapor; noise-equivalent power of a detector

Z zenith; upper boundary of atmosphere; angle in a pyrheliometer collimator

a area; absorptance; a constant; superscript indicating absorption; major axis of ellipse; a Mie coefficient

b width; a constant; line-shape factor; minor axis of ellipse; a Mie coefficient

c speed of light; indicator for cloud; a constant

d distance; mean distance between absorption lines; indicator for diffuse

e base of natural logarithms; emission coefficient; unit electrical charge; water vapor pressure; indicator for emission; label for coordinate axis

f a function; flux; luminousity function

g acceleration of gravity

h Planck's constant; hour angle

i a numerical index; electric current

j a numerical index; emission coefficient

k a numerical index; a constant; Boltzmann's constant; absorption coefficient

l a numerical index; indicator for absorption line

m a numerical index; mass; air mass; index of refraction

n a numerical index; index of refraction

p pressure; a point; phase function; coefficient of a Fourier series

q a flux of energy; coefficient of a Fourier series

r distance or length; electrical resistance; particle radius; diffusivity or diffusion factor; label for coordinate axis

s distance or length; indicator for scattering

t time; variable optical thickness

u optical mass or pathlength of a gas

v electromotive force; instrument signal; amplitudes of harmonic components

w width

x quantity (2π) (particle radius/wavelength)

y quantity (x) (index of refraction)

z height or altitude

Δ a small quantity; a difference; depolarization factor

Θ scattering angle

π angular function in Mie theory

T angular function in Mie theory

Φ function in Curtis–Godson approximation

χ angle of plane of polarization with respect to e-axis

Ψ function in Curtis–Godson approximation

Ω solid angle

α a constant; an angle; Ångström wavelength exponent; half-width of absorption line; absorptance of a sensor

β Ångström turbidity coefficient; absorptance of a sensor; measure of ellipticity

γ an angle; thermal conductivity

δ phase angle; declination of the Sun

ϵ emissivity

ξ angle of plane of transmission of polarizer; Riccati–Bessel function

θ an angle; angle from zenith or nadir directions

κ coefficient of absorption, emission, or attenuation

λ wavelength

μ cosine of angle from zenith or nadir $(\mu = \cos\theta)$

ν frequency

ν^{*} slope of curve for Junge aerosol distribution

ρ density

τ optical thickness

ϕ azimuth; latitude; probability integral; instrument response; work function of a material

ψ an angle; Riccati-Bessel function; effective aperture or angle factor of a receiver

ω an angle; solid angle; rotational frequency

$\parallel$ symbol indicating parallel

$\perp$ symbol indicating perpendicular

$*$ superscript representing surface reflection

Relative Sensitivity vs Wavelength of the Normal Light-Adapted Human Eye[a]

Wavelength (μm)	Relative sensitivity	Wavelength (μm)	Relative sensitivity
0.38	0.0000	0.57	0.9520
0.39	0.0001	0.58	0.8700
0.40	0.0004	0.59	0.7570
0.41	0.0012	0.60	0.6310
0.42	0.0040	0.61	0.5030
0.43	0.0116	0.62	0.3810
0.44	0.0230	0.63	0.2650
0.45	0.0380	0.64	0.1750
0.46	0.0600	0.65	0.1070
0.47	0.0910	0.66	0.0610
0.48	0.1390	0.67	0.0320
0.49	0.2080	0.68	0.0170
0.50	0.3230	0.69	0.0082
0.51	0.5030	0.70	0.0041
0.52	0.7100	0.71	0.0021
0.53	0.8620	0.72	0.0010
0.54	0.9540	0.73	0.0005
0.55	0.9950	0.74	0.0003
(0.55)	1.0002	0.75	0.0001
0.56	0.9950	0.76	0.0001

[a] Adopted by Commission Internationale de L'Eclairage in 1931; Condon and Odishaw, 1958.

Spectral Distribution of Solar Radiation
Incident at the Top of the Atmosphere for a
Mean Sun–Earth Distance as Adopted by
NASA as a Standard for Purposes of
Engineering Design[a]

λ [b] (μm)	H_λ [c] (W cm^{-2} μm^{-1})	P [d] (%)
0.120	0.000010	0.00044
0.140	0.000003	0.00054
0.150	0.000007	0.00058
0.160	0.000023	0.00069
0.170	0.000063	0.00101
0.180	0.000125	0.00170
0.190	0.000271	0.00316
0.200	0.00107	0.0081
0.210	0.00229	0.0205
0.220	0.00575	0.0502
0.225	0.00649	0.0729
0.230	0.00667	0.0972
0.235	0.00593	0.1205
0.240	0.00630	0.1430
0.245	0.00723	0.1681
0.250	0.00704	0.1944
0.255	0.0104	0.2267
0.260	0.0130	0.270

λ [b] (μm)	H_λ [c] (W cm^{-2} μm^{-1})	P [d] (%)
0.265	0.0185	0.328
0.270	0.0232	0.405
0.275	0.0204	0.486
0.280	0.0222	0.564
0.285	0.0315	0.644
0.290	0.0482	0.811
0.295	0.0584	1.008
0.300	0.0514	1.211
0.305	0.0603	1.417
0.310	0.0689	1.656
0.315	0.0764	1.924
0.320	0.0830	2.219
0.325	0.0975	2.552
0.330	0.1059	2.928
0.335	0.1081	3.324
0.340	0.1074	3.722
0.345	0.1069	4.118
0.350	0.1093	4.517
0.355	0.1083	4.919
0.360	0.1068	5.317
0.365	0.1132	5.723
0.370	0.1181	6.151
0.375	0.1157	6.583
0.380	0.1120	7.003
0.385	0.1098	7.413
0.390	0.1098	7.819
0.395	0.1189	8.242
0.400	0.1429	8.725
0.405	0.1644	9.293
0.410	0.1751	9.920
0.415	0.1774	10.57
0.420	0.1747	11.22
0.425	0.1693	11.86
0.430	0.1639	12.47
0.435	0.1663	13.08
0.440	0.1810	13.73
0.445	0.1922	14.42
0.450	0.2006	15.14
0.455	0.2057	15.89
0.460	0.2066	16.65
0.465	0.2048	17.41
0.470	0.2033	18.17
0.475	0.2044	18.92

λ [b] (μm)	H_λ [c] (W cm^{-2} μm^{-1})	P [d] (%)
0.480	0.2074	19.68
0.485	0.1976	20.43
0.490	0.1950	21.16
0.495	0.1960	21.88
0.500	0.1942	22.60
0.505	0.1920	23.31
0.510	0.1882	24.02
0.515	0.1833	24.70
0.520	0.1833	25.38
0.525	0.1852	26.06
0.530	0.1842	26.74
0.535	0.1818	27.42
0.540	0.1783	28.08
0.545	0.1754	28.74
0.550	0.1725	29.38
0.555	0.1720	30.02
0.560	0.1695	30.65
0.565	0.1705	31.28
0.570	0.1712	31.91
0.575	0.1719	32.54
0.580	0.1715	33.18
0.585	0.1712	33.81
0.590	0.1700	34.44
0.595	0.1682	35.06
0.600	0.1666	35.68
0.605	0.1647	36.30
0.610	0.1635	36.90
0.620	0.1602	38.10
0.630	0.1570	39.27
0.64	0.1544	40.42
0.65	0.1511	41.55
0.66	0.1486	42.66
0.67	0.1456	43.74
0.68	0.1427	44.81
0.69	0.1402	45.86
0.70	0.1369	46.88
0.71	0.1344	47.88
0.72	0.1314	48.86
0.73	0.1290	49.83
0.74	0.1260	50.77
0.75	0.1235	51.69
0.80	0.1107	56.02
0.85	0.0988	59.89

λ [b] (μm)	H_λ [c] (W cm^{-2} μm^{-1})	P [d] (%)
0.90	0.0889	63.36
0.95	0.0835	66.54
1.0	0.0746	69.46
1.1	0.0592	74.41
1.2	0.0484	78.39
1.3	0.0396	81.64
1.4	0.0336	84.34
1.5	0.0287	86.64
1.6	0.0244	88.61
1.7	0.0202	90.26
1.8	0.0159	91.59
1.9	0.0126	92.64
2.0	0.0103	93.49
2.1	0.0090	94.20
2.2	0.0079	94.83
2.3	0.0068	95.37
2.4	0.0064	95.86
2.5	0.0054	96.29
2.6	0.0048	96.67
2.7	0.0043	97.01
2.8	0.0039	97.31
2.9	0.0035	97.58
3.0	0.0031	97.83
3.1	0.0026	98.04
3.2	0.00226	98.22
3.3	0.00192	98.37
3.4	0.00166	98.50
3.5	0.00146	98.62
3.6	0.00135	98.72
3.7	0.00123	98.82
3.8	0.00111	98.91
3.9	0.00103	98.98
4.0	0.00095	99.06
4.1	0.00087	99.13
4.2	0.00078	99.19
4.3	0.00071	99.24
4.4	0.00065	99.29
4.5	0.00059	99.34
4.6	0.00053	99.38
4.7	0.00048	99.42
4.8	0.00045	99.45
4.9	0.00041	99.48
5.0	0.00038	99.51

λ [b] (μm)	H_λ [c] (W cm^{-2} μm^{-1})	P [d] (%)
6.0	0.00018	99.71
7.0	0.00010	99.82
8.0	0.00006	99.88

[a] After Thekaekara, 1971.

[b] λ: wavelength in μm.

[c] H_λ: irradiance in a narrow band centered on λ in W cm^{-2} m^{-1}.

[d] P: percent of solar constant associated with wavelengths shorter than λ.

Index

A
B   5
C   6
D   7
E   8
F   9
G   0
H   1
I   2
J   3